Aging and Recovery of Function in the Central Nervous System

Aging and Recovery of Function in the Central Nervous System

Edited by

Stephen W. Scheff

University of Kentucky
College of Medicine
Lexington, Kentucky

Plenum Press • *New York and London*

Library of Congress Cataloging in Publication Data

Main entry under title:

Aging and recovery of function in the central nervous system.

Includes bibliographical references and index.
1. Brain damage—Age factors. 2. Brain damage—Animal models. 3. Central nervous systems—Diseases—Age factors. 4. Central nervous system—Aging. I. Scheff, Stephen W. [DNLM: 1. Brain injuries—Physiopathology. 2. Brain—Physiopathology. 3. Aging. WL 354 A267]
RC387.5.A45   1984   618.97'68307   83-26900
ISBN-13: 978-1-4612-9680-5   e-ISBN-13: 978-1-4613-2705-9
DOI: 10.1007/978-1-4613-2705-9

© 1984 Plenum Press, New York
Softcover reprint of the hardcover 1st edition 1984
A Division of Plenum Publishing Corporation
233 Spring Street, New York, N.Y. 10013

Contributors

C. ROBERT ALMLI   Departments of Preventive Medicine, Anatomy and Neurobiology, and Psychology, Programs in Occupational Therapy and Neural Sciences, Washington University School of Medicine, St. Louis, Missouri 63110

KEVIN ANDERSON   Department of Anatomy, University of Kentucky College of Medicine, Lexington, Kentucky 40536

ARTHUR L. BENTON   Departments of Neurology and Psychology, University of Iowa, Iowa City, Iowa 52240

HERMAN BLUMENTHAL   Department of Psychology, Washington University, St. Louis, Missouri 63130

JAMES R. CONNOR   Department of Physiology–Anatomy, University of California, Berkeley, California 94720

JAMES V. CORWIN   Department of Psychology, University of Kentucky, Lexington, Kentucky 40506. *Present address:* Department of Neurology, University of Florida Medical Center, Gainesville, Florida 32610

STEPHEN D. CURTIS   Department of Psychology, University of Kentucky, Lexington, Kentucky 40506. *Present address:* Psychology Department, Indiana University, Bloomington, Indiana 47401

STEVEN T. DeKOSKY   Department of Neurology, Lexington Veterans Administration and University of Kentucky Medical Centers and Sanders–Brown Research Center on Aging, Lexington, Kentucky 40536

MARIAN C. DIAMOND   Department of Physiology–Anatomy, University of California, Berkeley, California 94720

LINDA EWING-COBBS   Department of Psychology, University of Houston, Houston, Texas 77004

STANLEY FINGER   Department of Psychology and Neurobiology Program, Washington University, St. Louis, Missouri 63130

WALTER L. ISAAC   Department of Psychology, University of Kentucky, Lexington, Kentucky 40506

HARVEY S. LEVIN   Division of Neurosurgery, The University of Texas Medical Branch, Galveston, Texas 77550

JOHN F. MARSHALL   Department of Psychobiology, University of California, Irvine, Irvine, California 92717

ARTHUR J. NONNEMAN   Department of Psychology, University of Kentucky, Lexington, Kentucky 40506

STEPHEN W. SCHEFF   Departments of Anatomy and Neurology, University of Kentucky College of Medicine, Lexington, Kentucky 40536

DONALD G. STEIN   Department of Psychology, Clark University, Worcester, Massachusetts 01610, and Department of Neurology, University of Massachusetts Medical Center, Worcester, Massachusetts 01605

M. LISA VALENTINO   Department of Psychology, Clark University, Worcester, Massachusetts 01610

JOHN P. VICEDOMINI   Department of Psychology, University of Kentucky, Lexington, Kentucky 40506

RICHARD F. WALKER   Department of Anatomy and Sanders–Brown Research Center on Aging, University of Kentucky Medical Center, Lexington, Kentucky 40536

DAVID WOZNIAK   Department of Psychology, Washington University, St. Louis, Missouri 63130

Preface

The mammalian central nervous system is a remarkable structure which has attracted many new investigators, as evidenced by the dramatic increase in scientific publications dealing with neurobiology. Every day basic scientists conduct new and exciting experiments, resulting in remarkable discoveries destined to help mankind. Unfortunately many of these new findings are slow to be accepted by the clinical world. This is especially true in the area of brain trauma, where the present prognosis is usually very poor. We have known for quite some time that the mammalian central nervous system is capable of compensating for severe damage in many different ways, and under some circumstances behavioral compensation can be observed. However, much is still to be learned about the various factors and events that lead to functional recovery and those conditions that do not. It is this challenge that originally excited a number of the contributors to this volume to explore the subject of recovery from brain damage.

One factor in particular that is known to change the prognosis of recovery is the age of the organism at the time of the damage. This book is an attempt to explore this important variable. Most of the literature concerning aging deals with widespread degenerative changes and paints a grim picture for the aging central nervous system in terms of recovery of function following trauma. One of the problems associated with aging research in general stems from the realization that chronological age does not necessarily equal biological age. The aging process is very complex and it is not enough to simply study qualitative physiological and morphological changes in old animals and people. While it is known that a wide spectrum of changes occur in neurons and the neuropil with advancing age, it is also known that the age-related changes are not characteristic of all parts of the nervous system. It is important to know under what conditions the young and aged nervous systems are alike and different. In many cases it appears that the aged organism performs as well as its younger counterpart. In other instances, such as under conditions of stress, the aged central nervous system begins to perform differently. This book then is an attempt to integrate these two different yet fascinating fields of neurobiology: aging and recovery of function. Many of the authors have been working in the field of recovery of function for many years. I have asked each to evaluate a specific sector of the problem using his own experimentation and that of others to illustrate important points.

It is hoped that this book will be of interest to a broad audience, including basic scientists, their students, and clinicians. It is also my hope that this volume will stimulate new research in an exciting field of neurobiology, that dealing with recovery from brain damage and the factor of aging. We have only begun to explore the many different facets of this important problem.

Stephen W. Scheff

Lexington, Kentucky

Contents

Chapter 3

Morphological Measurements in the Aging Rat Cerebral Cortex

MARIAN C. DIAMOND AND JAMES R. CONNOR

Chapter 4

Morphologic Aspects of Brain Damage in Aging

STEPHEN W. SCHEFF, KEVIN ANDERSON, AND
STEVEN T. DEKOSKY

Chapter 8

Age, Brain Damage, and Behavioral Recovery

ARTHUR J. NONNEMAN, JOHN P. VICEDOMINI,
JAMES V. CORWIN, STEPHEN D. CURTIS, AND
WALTER L. ISAAC

Chapter 9

Age and Recovery from Brain Damage: A Review of Clinical Studies

HARVEY S. LEVIN, LINDA EWING-COBBS, AND
ARTHUR L. BENTON

Chapter 10
Recovery of Function in Senile Dementia of the Alzheimer Type
STEVEN T. DEKOSKY

1

Longevity, Disease, and Autoimmune Reactions following Focal Cortical Injuries

STANLEY FINGER, DAVID WOZNIAK,
and HERMAN BLUMENTHAL

1. INTRODUCTION

The relationship between brain damage and the aging process is not well under-stood. On the one hand, it is clear that irradiation restricted to the brain can markedly shorten lifespan (Ordy *et al.*, 1967; Samorajski *et al.*, 1970; Rolsten *et al.*, 1973). But, on the other hand, the brain changes resulting from irradiation procedures are widespread and shed little light upon the contributions of different brain areas to aging and age-related pathology. Moreover, while there have been a few studies comparing the effects of focal brain damage in sexually mature and very old animals (e.g., Stein and Firl, 1976), these experiments have essentially been concerned with cognitive, motivational, and emotional changes and not with factors such as morbidity and longevity.

Nevertheless, the presumption among workers in this field seems to be that some parts of the brain may be more critically involved in the aging process than others. In particular, the hypothalamus is often singled out as a possible mediator of aging and age-related pathology because many metabolic and behavioral changes that are seen with aging seem to involve the hypothalamic–neuroendo-crine axis. Furthermore, hypothalamic lesions in relatively young laboratory animals result in sexual, emotional, metabolic, and appetitive changes that bear

STANLEY FINGER • Department of Psychology and Neurobiology Program, Washington University, St. Louis, Missouri 63130. DAVID WOZNIAK and HERMAN BLUMENTHAL • Department of Psychology, Washington University, St. Louis, Missouri 63130.

some resemblance to those seen later in life (Almli, Chapter 2). Whether aging and age-related pathology represent increased or decreased hypothalamic functioning, however, is still a matter of debate (Dilman, 1976, 1979; Groen, 1959; Frolkis *et al.*, 1972; Frolkis, 1976), and whether some of the proposed hypothalamic changes actually represent cause or effect is, to say the least, not always clear.

The sparsity of actual aging studies on the hypothalamus may be cited as a major reason for the diverse opinions about the role that this structure may play in the aging process. Still, it can be stated that even less is known about the relationships between other brain areas and aging and how damage to higher parts of the brain may affect lifespan. Patients with arterial disease and various forms of chronic brain pathology obviously may not live as long as individuals without brain disease, but what about seemingly healthy individuals who receive cortical injuries from an external source and who appear to show good or perhaps even complete recovery? Would such individuals be expected to have normal lifespans, and would these patients be expected to die of the same diseases as members of the non-brain-damaged population?

Because so little is known about the long-term effects of acute brain injuries in otherwise healthy individuals, we have recently conducted a series of controlled experiments using the C57 laboratory mouse as a model. These studies were designed to learn more about the interaction between acute cortical lesions and lifespan, and how morbidity and immunological responses might also be affected by focal brain injuries from an external source. The most significant features and results of these investigations will be summarized here. Wozniak *et al.* (1982) and Blumenthal *et al.* (1984) should be consulted for more detailed descriptions of these experiments.

2. LONGEVITY

Knowing whether a healthy person who experiences an industrial accident, a gunshot wound, or even a brain biopsy would be expected to have a shortened lifespan is important for many reasons. Data collected on such individuals might shed light on the concept of an aging pacemaker in the brain (Denkla, 1977; Dilman, 1976; Finch, 1976; Timiras, 1978) and the idea that some diseases may reflect defects in such a regulator of the aging process. Such data would also bear directly on decisions to perform brain biopsies, psychosurgery, and related operations which involve damaging the blood–brain barrier and the possibility of subsequent autoimmune reactions (Threatt *et al.*, 1971). This type of information would also be important to scientists wishing to study the long-term effects of brain lesions on behavior, anatomy, or physiology. Here, it is important to know whether subjects in the experimental or treatment groups would be available for study months or years after damage, or whether shorter-term experiments should be planned. And, should lifespan be affected by brain damage, research along these lines may suggest interventions to normalize it.

With these considerations in mind we (Wozniak *et al.*, 1982) subjected over 300 male C57BL/6J mice (Jackson Laboratories) to either bilateral frontal cortex transections or sham operations when they were 64–82 days of age. The lesions were performed under methoxyflurane anesthesia (Penthrane, Abbott), and involved transversely inserting a pointed, No. 11 scalpel blade 1.5 mm into the brain approximately 1.5 mm behind the anterior limits of the frontal cortex. The blade was then arced in a medial-lateral plane to produce a long coronal cut on each side of the brain (see Glick *et al.*, 1971). Almost all mice survived this procedure or the sham operation (which involved the cutting of the fascia to expose the skull but not brain damage), and were housed in clear plastic cages (29 × 19 × 13 cm) for the rest of the experiment. The cages originally contained eight animals each, but this number diminished as animals died or were sacrificed over the course of the investigation.

Because stress has also been shown to affect longevity, in some cases by lengthening it and in others by shortening it (Sacher, 1977), approximately half of the animals in each group were intermittently exposed to cold stress. This involved removing food and water from their cages and placing them in a ventilated cold (10°C) chamber for 3 hr. The lesion and control animals experienced this stress when 2½, 9, 16, 20, and 25 months of age.

Following the 9- and 25-month cold-stress sessions, five animals from the stressed and nonstressed lesion and control groups were sacrificed to evaluate the immediate effects of the cold stressor on stomach ulceration and serum corticosterone levels, two frequently used indicators of stress, and to assess the general condition of the subjects in each treatment group (see Poland *et al.*, 1981, and Wozniak and Goldstein, 1980, for corticosterone and gastric erosion evaluation procedures, respectively). In addition, five animals from each of the four groups were sacrificed when 30 months of age (i.e., 5 months after some last-experienced cold stress) in order to see whether the effects of stress would still be noticeable after a long recovery period and to determine how disease incidence was changing in old age. The animals not sacrificed were allowed to live out their lives in the cages, and records were kept of when each died. In most cases it was possible to retrieve the bodies of the animals soon after they died, and in these instances pathological studies were performed to assess the probable cause of death.

That the cold procedure was in fact stressful was clear from the serum corticosterone data. The animals experiencing cold stress exhibited a mean level of about 12 μg/dl, as compared with the nonstressed mice whose mean level was about 2 μg/dl. Analyses of variance showed that these differences were highly significant. Basal resting levels 5 months after the last cold-stress session, however, were comparable for previously stressed and nonstressed mice. In no instance was there any evidence that the frontal lesions affected the corticosterone levels; the only significant findings pertained to whether the animals were stressed or not immediately before being sacrificed. Nevertheless, in contrast to the serum corticosterone results, stomach ulceration was little affected by the cold stress ($Ps > 0.05$). Brain damage also had no effect on this dependent variable.

The longevity data for the mice that died from "natural causes" are shown in Fig. 1. As can be seen, the lifespan profiles for the four groups were very similar, regardless of whether the mice experienced sham operations or brain lesions, stress or no stress. Very few animals died at less than 2 years of age, and even fewer lived to be 3 years old. The group means ranged from 27.9 months to 29.8 months, with standard deviations averaging approximately 5 months. An analysis of variance was conducted on these data and it confirmed that there were no significant main or interaction effects ($Ps > 0.05$). These results indicate that

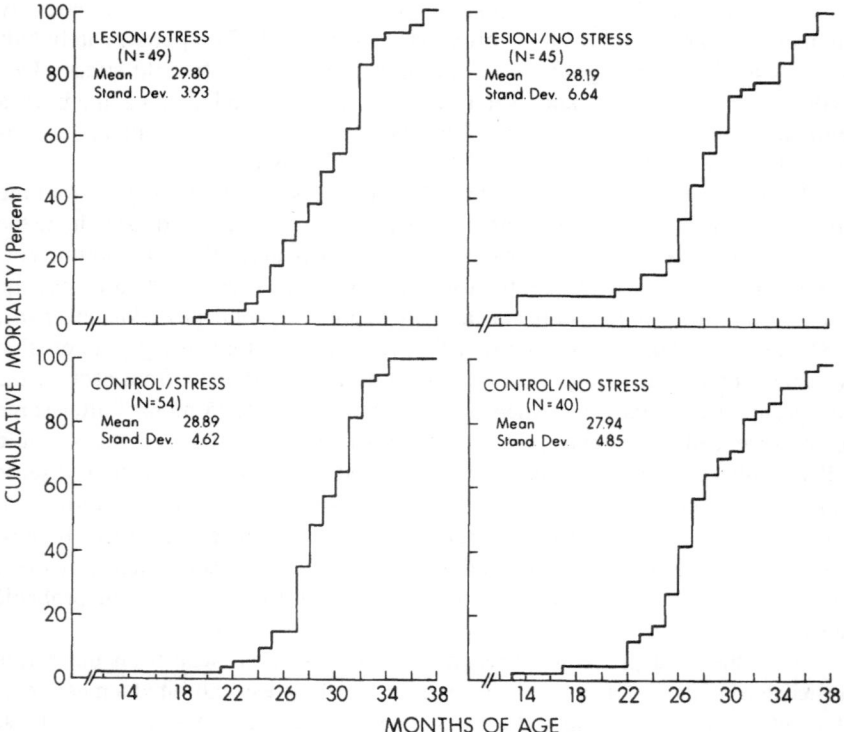

Figure 1. Cumulative mortality rates as a function of surgical and stress treatments. [From Wozniak *et al.* (1982), reproduced with permission of the *Journal of Gerontology.*]

Figure 2. Sagittal section of mouse brain showing frontal pole transection (arrow). These lesions were consistent from animal to animal and extended from within 1 mm of the midline of the lateral edge of the cortex just anterior to bregma.

lifespan is not necessarily affected when a previously healthy subject suffers an acute, focal cortical wound.

Figure 2 shows a section of mouse brain revealing a representative frontal lesion. These lesions typically penetrated as deeply as the corpus callosum and extended from the midline to the lateral surface of the frontal cortex on each side of the brain. In all animals studied, the cuts were found to be placed as intended.

In summary, the control animals in this investigation showed essentially normal lifespans and proved to be comparable to unoperated mice described in other experiments (e.g., Rowlatt *et al.*, 1976, reported a mean longevity score of 26.6 months, as compared with 27.9 months for the control group in the present study). Acute, focal brain lesions and/or stress did not result in any obvious behavioral changes in the animals and did not alter longevity.

These findings can be contrasted with the results of experiments conducted by Ordy *et al.* (1967), Samorajski *et al.* (1970), and Rolsten *et al.* (1973). These investigators, working as a team, found that brain damage can significantly shorten lifespan in the mouse. Their procedure for damaging the brain, however, differed markedly from that used here. The whole brain was exposed to deuteron or proton irradiation in these studies. This resulted in a loss of neural cells, glial proliferation, and vascular changes which were not restricted to a single locus (see Samorajski *et al.*, 1970). Thus, while showing that the brain can play a role in the regulation of the aging process, the irradiation techniques shed little light on the contributions of specific brain areas to longevity and differ in other dimensions from the surgical lesion procedure used here.

As for the possibility that the brain may actually contain a pacemaker for the rate of aging (Timiras, 1978), it should be pointed out that the evidence for this is indirect and equivocal at present, with interest largely centered around the hypo-thalamic–pituitary segment of the neuroendocrine system (Dilman, 1976; Finch, 1976; Timiras, 1978). At any rate, although the frontal cortex has connections with the hypothalamus, it does not appear as if the frontal cortex transections are affecting the aging process. Of course, this does not prove that the frontal cortex is not directly involved in aging since it may play a role in the more cognitive changes associated with aging, or that higher centers cannot influence the work-ing of a structure such as the hypothalamus. Regarding the latter possibility, if the frontal cortex is part of a diffuse and redundant network that can affect aging, transections like those made here would be expected to have minimal effects. Large ablations rather than transections of the frontal cortex, as well as lesions of other areas of the brain, may shed additional light on the possibility of a cortical contribution to the aging process, whether direct or indirect in nature.

Finally, it should be mentioned that the observation that cold stress did not affect lifespan was not entirely unexpected. Although there may be a progres-sive, age-related decline in the capacity to reestablish equilibrium conditions following a major stressful event, the literature is contradictory and in some instances stressors such as cold, starvation, or electric shock have even been found to extend longevity (see Sacher, 1977; Liu and Walford, 1972). Thus, many conditions and factors could determine the results of experiments on stress and longevity, such as the duration and intensity of the stressors, age at the time of exposure, and the species and strain of animal being studied.

3. DISEASE PROFILES

Whether brain damage affects longevity, and whether it affects the probable cause of death, are two different questions. That is, even though longevity differences were not observed following frontal cortical damage, it is conceiv-able that subjects with brain lesions are more likely to die from certain diseases (e.g., more degenerative neural diseases) than those without brain damage. For this reason, pathological examinations were performed whenever possible on the animals that died in the longevity study just described. These examinations involved all major organ systems and were specifically aimed at determining the most probable cause of death. Similar examinations were performed on the animals that were sacrificed at various points in the study with the hope of detecting the early presence of diseases that could have affected their lifespans.

The examinations revealed that the principal causes of death in the study were pulmonary adenomas, metastatic pulmonary tumors, reticulum cell sarcomas, leukemias, and pneumonias. The pneumonias were generally of the bronchial type, but sometimes involved a whole lobe of the lung. Although the pneumonias occasionally appeared alone, they frequently were seen in the presence of other diseases, in particular, pulmonary adenomas. When this occurred, the pneumonia was considered to be secondary, and the adenoma was recorded as the probable cause of death. Although the adenomas of the lung were histologically benign, these tumors frequently were large and multiple at the time of death, sometimes severely obstructing the pulmonary tree and replacing lung parenchyma.

The reticulum cell sarcomas were frequently seen in the spleen which then became markedly enlarged. These sarcomas often spread to the liver which also became massive. Some mice with reticulum cell sarcomas showed metastasis to the lungs. In the latter cases, the probable cause of death was recorded as metastatic pulmonary tumors.

Leukemias were observed, and these generally infiltrated the liver and the spleen. The pathological examinations also revealed a few mice with amyloid of the renal glomeruli and lungs, and a few with granulomatous myocarditis.

Table 1 shows the frequencies of the various diseases judged to cause death as a function of brain damage and stress for the animals described in the longevity study. These data were analyzed with a multiple frequency table analysis (see Wozniak et al., 1982) which employed a long-linear model (Fienberg, 1979). This analysis revealed a significant relationship (partial association) between stress and cause of death, but not between brain damage and cause of death.

Table 1. Frequencies of Diseases Associated with Natural Death[a]

	Group			
Disease	Sham operation/ stress	Lesion/ stress	Sham operation/ no stress	Lesion/ no stress
Leukemia	17	12	16	10
Pneumonia	14	13	8	10
Reticulum cell sarcoma	4	0	5	5
Metastatic pulmonary tumor	5	3	0	0
Pulmonary adenoma	2	4	0	4
Other	7	13	7	6
TOTAL	49	45	36	34

[a]After Wozniak et al. (1982), reproduced with permission of the *Journal of Gerontology*.

Further analyses on the mice that died from natural causes showed that two diseases were major contributors to this effect: reticulum cell sarcoma ($P <$ 0.032) and metastatic pulmonary tumors ($P < 0.002$). Metastatic pulmonary tumors were positively related to stress, whereas the reticulum cell sarcomas were negatively related to stress.

The animals sacrificed throughout the experiment showed very little evidence of disease prior to the second year of life. Thereafter, their disease profiles resembled those of mice that died from natural causes, with the possible exception of a trend for stressed, sacrificed mice to show pulmonary adenomas. Sample sizes and frequencies, however, were too small for statistical treatment of the data.

These results suggest that frontal cortex damage does not affect longevity or morbidity, or result in different pathological profiles at the time of death, at least under the conditions of this study. Nevertheless, stress, which also did not seem to affect lifespan, appeared to affect disease incidence and probable cause of death. Specifically, intermittent exposure to cold stress increased the probability of seeing pulmonary tumors. However, in the case of the metastatic pulmonary tumors, it should be remembered that these tumors originated from reticulum cell sarcomas. Thus, among the mice that died from natural causes, the negative relationship found between stress and reticulum cell sarcomas as the single most likely cause of death should not be interpreted to mean that reticulum cell sarcomas were not found in this sample.

With regard to the *positive* relationship between stress and metastatic pulmonary tumors, the idea that stress can lead to a faster spread of malignancy is relatively well known from both clinical and laboratory animal studies. Sklar and Anisman (1979), for example, reported that the growth of experimentally induced mastocytomas in male mice was enhanced by exposing the animals to a stressful situation in which they were unable to execute effective coping responses. In the aforementioned study, a single stress session with inescapable shock produced an earlier appearance of the tumor, an enlargement of its size, and shorter survival times. The data collected here support this type of finding.

4. NEURON-BINDING ANTIBODIES

The idea that there may be antibodies that bind with neurons is now attracting considerable attention. Neuron-binding antibodies have recently been associated with a number of neurological and psychiatric disorders, including myasthenia gravis (Martin *et al.*, 1974), carcinomatous neuropathy (Zeromski, 1970), and perhaps even schizophrenia (Heath and Krupp, 1967; Kolyaskina and Kushner, 1969). Antibodies with the potential to bind with neurons also seem to appear in the blood of seemingly healthy mice (Threatt *et al.*, 1971; Nandy,

1975, 1977), rats (Miller and Blumenthal, 1978), monkeys (Felsenfeld and Wolf, 1972), and humans (Felsenfeld and Wolf, 1972; Ingram *et al.*, 1974) as a function of the aging process. For example, Nandy (1972) has reported that these antibodies first appear in the blood of C57 mice when the animals are between 6 and 15 months of age.

Although Threatt *et al.* (1971) identified antibodies that could bind with neurons in the *serum* of older mice, these investigators noted only minimal binding within the brain itself as a function of age. They proposed that the blood–brain barrier normally serves to prevent serological antibodies from reacting with brain tissue, and that the immunoglobulins could be cytotoxic, i.e., that they might represent one mechanism that could account for the progressive diminution of brain cells over the course of a lifetime (Brody, 1955; Vogel, 1969). Nevertheless, whether the blood–brain barrier actually weakens with the passage of time still is not clear (Feden *et al.*, 1979), and the functional significance of the antibodies associated with aging is currently a matter of debate.

In contrast to the new but growing literature on neuron-binding antibodies in later life, the immunological response to acute brain damage has hardly been explored. In the Soviet Union, Motyka and Jezkova (1975) and Savenko and Polienko (1975) have reported that levels of brain-reactive antibodies may be elevated by strokes and disorders of brain circulation in their patients. However, the temporal parameters of these phenomena and the history of brain pathology prior to hospitalization are difficult to determine with patient populations. In the laboratory, Threatt *et al.* (1971) found that inserting a chilled metal probe into the brains of mice had no effect on antibody binding if the mice were under 6 months of age, although a reaction was seen in the vicinity of the lesion in older mice. The mice in this experiment were sacrificed only 6–12 hr after injury and, on the basis of other data collected by these investigators, it seems reasonable to conclude that the reaction seen in the older animals was due to antibodies already in circulation that normally would not have penetrated the blood–brain barrier in large numbers.

Because so little is known about the antibody response following physical damage to the blood–brain barrier, the mice described in the preceding experiments on longevity and morbidity were examined for IgM and IgG nuclear and cytoplasmic binding to neurons with the "direct" immunoperoxidase procedure described by one of us in an earlier study (Miller and Blumenthal, 1978). The analyses utilized horseradish peroxidase-tagged rabbit antimouse immunoglobulins (RAM IgG and IgM), and assessed the binding that took place while the mice were still alive. The intensity of this binding was subjectively graded on the basis of color (which ranged from light tan to dark brown or black) with a "blind" procedure using a 0 to 4+ (4.0) scale. As in earlier work from this laboratory, a score of 2+ signified a "positive" binding reaction (see Miller and Blumenthal, 1978) (Fig. 3). The score for each animal was based on the most

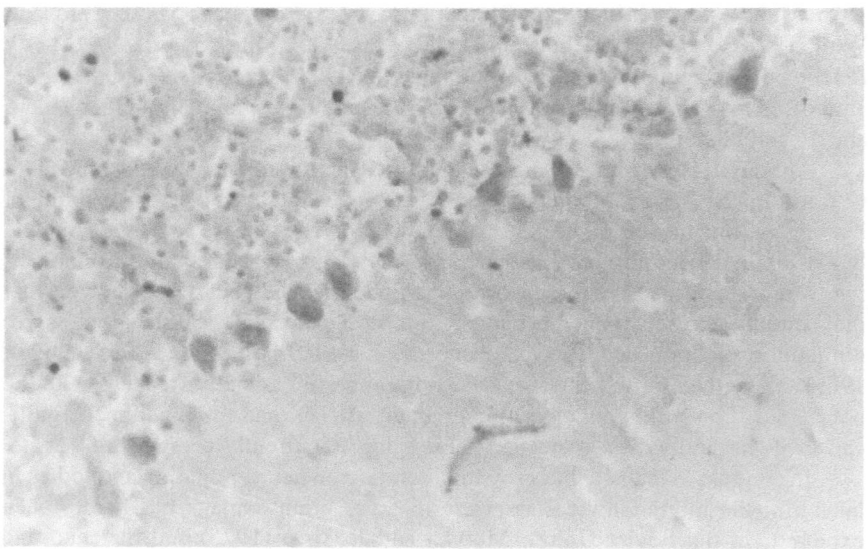

Figure 3. Section of cerebellum of a mouse showing IgM binding of Purkinje cells. Some cells show both nuclear and cytoplasmic binding. Magnification approximately 250×.

intense binding noted in the sample of brain tissue coming from the general vicinity of the lesion or from a comparable location in control animals. Tests for specificity of nuclear binding (e.g., binding to nuclei of cells of pancreas, kidney, and liver) in positive brain cases and a variety of control procedures (use of peroxidase-tagged albumin; tests of endogenous peroxidase levels involving untagged antiserum) were incorporated to comprehend more fully the immunological reaction to acute brain injury and/or repeated stress.

We (Blumenthal *et al.*, 1984) first looked at the mice that were allowed to die from natural causes. These data are shown in Fig. 4. The binding that was noted was almost exclusively neuronal, and with only a few exceptions, nuclear as opposed to cytoplasmic. The mean levels for the groups were below 2+, however, even after the groups were subdivided by a median split into those that lived under 30 months and those that lived for 30 or more months. Furthermore, whenever there was a positive nuclear-binding reaction to neurons, there was also a correponding nuclear reaction in one or more of the other test organs (kidney, liver, and pancreas).

Since the major concern of this study was to see if the number of cases showing significant binding to neurons (scores of 2+ or more) increased as a function of brain damage, stress, or age, the dichotomized data (less than 2+ versus 2+ or greater reactions) were analyzed (where possible) with multiple frequency table analyses (Fienberg, 1979) using a log-linear model.

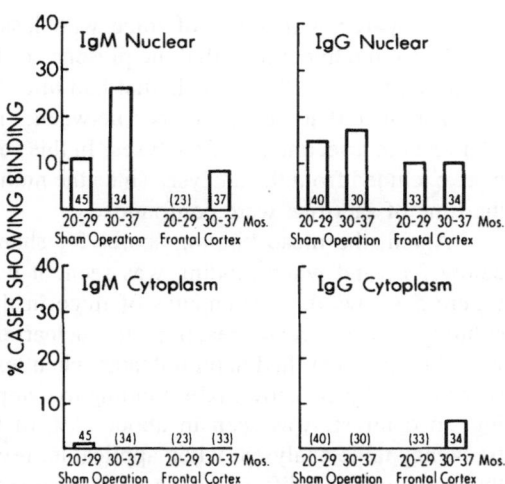

Figure 4. Percent of mice dying from natural causes that showed binding scores of 2+ or greater to each of the four types of immunoglobulin binding. Animals with lesions and those with sham operations are subdivided into those that lived more than or less than 2½ years. Stressed and non-stressed groups were pooled for the construction of this figure since their binding profiles were alike and since no effects of stress emerged in the statistics on those cases showing binding responses. The number at the base of each bar signifies sample size.

In the case of the IgM nuclear data, this statistic showed that the best fitting, most parsimonious model was $(LA) (BA) (BL) (BDS)$, where A = age, S = stress, L = lesion, D = cause of death, and B = antibody score. Stepwise deletion procedures revealed four significant partial correlations. One (BA) related to age $(P < 0.025)$, and showed that IgM nuclear binding did in fact increase in animals that lived to be at least 2½ years of age. The second (BL) related to the lesion effect $(P < 0.01)$ and, contrary to our expectations, indicated that IgM nuclear binding was more likely to be seen after a sham operation than after a cortical lesion, at least given these lengthy recovery periods. The third significant partial correlation (LA) reflected more brain-damaged mice in the ≥30-month sample than in the <30-month sample. However, this effect must be viewed in the context of the Wozniak *et al.* (1982) study, which showed that the populations from which these samples were obtained did not differ in longevity. The last partial correlation $(BDS; P < 0.025)$—which showed a relationship among stress, positive binding scores, and cause of death—was difficult to interpret because it was largely due to our "Other" category, which included diseases with low frequencies of occurrence as well as deaths of unknown cause. Whether the two most frequent causes of death (pneumonia, leukemia) contributed significantly to this effect could not be determined due to zero marginals in the contingency table.

Multiple frequency table analyses of the IgG nuclear data did not reveal any significant partial correlations. In addition, there were too few mice showing 2+ or greater cytoplasmic-binding scores to IgG or IgM for statistics to be applied in these cases.

When the mean lifespans of the mice showing 2+ or greater binding scores were compared with those of mice with less intense reactions, there was no support for the hypothesis that the presence of binding may be related to longevity, at least among the animals that had already died from natural causes.

The mice that were sacrificed between 9 months and 2½ years of age were examined in a second set of analyses. In this case, however, the stressed animals were excluded from the analyses since the number of stress sessions endured and the time of sacrifice were confounded.

Again, the mean binding scores for sham-operated and lesion groups fell below 2+, and when binding was seen, it was virtually confined to neurons. Figure 5 shows the percentages of mice in the lesion and control groups that exhibited 2+ or greater reactions to nuclear and cytoplasmic IgM and IgG. As with the mice that died natural deaths when less than 2½ years of age, very few sacrificed subjects showed 2+ binding to cytoplasm. IgG neuronal nuclear binding, in contrast, was seen in about 25% of the mice. Nevertheless, multiple frequency table analyses, when applicable, revealed no evidence for an effect of the lesion or for a difference between mice sacrificed at 9–17 or 18–30 months of age on these measures.

Thus, the results of the analyses of antibody binding to neurons did not reveal any evidence for enhanced binding following earlier brain damage whether among the mice that died from natural causes or among those that were sacrificed at predetermined times by the experimenters. In fact, the data showed that IgM nuclear binding may even occur *less frequently* in the lesion group than in the control group if animals at least 2½ years of age are examined. Still, it must be recalled that all of these mice were allowed to live for at least 7 months

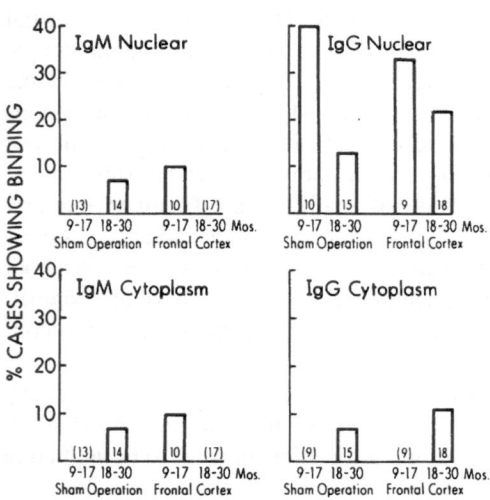

Figure 5. Percent of mice sacrificed that showed binding scores of 2+ or greater to each of the four types of immunoglobulin binding. The groups are subdivided into those sacrificed between 9 and 17 months and those sacrificed between 18 and 30 months of life. The number at the base of each bar signifies sample size.

after surgery, and that most lived for more than 1¼ years before being killed, or for more than 2¼ years before dying naturally. With this in mind, we decided to look at the possibility that there may be intense binding after a brain lesion, but that these effects are short-lived.

To test this hypothesis we prepared an additional ten C57BL/6 mice with frontal cortex lesions and ten more with sham operations, using the same techniques as before. One week after surgery, the animals (now approximately 100 days of age) were sacrificed. Again, the brains were sectioned and subjected to the direct immunoperoxidase procedures for assessing IgM and IgG nuclear and cytoplasmic binding. Since the data were unidimensional (no age factor), 2+ binding score frequencies were analyzed with Fisher's exact probability tests.

The neuronal data are presented in Fig. 6. As can be seen, there was no cytoplasmic binding evident in the control brains or at a distance from the lesion, but a clear IgG cytoplasmic reaction to neurons near to the lesion. The frequencies of the positive IgM and IgG neuronal nuclear reactions were also much higher near to the site of the lesion than distal to it or in the control brains and higher than had been seen previously. Moreover, with the short recovery period the animals with lesions now showed marked IgG and IgM glial binding (largely to cytoplasm) in the vicinity of the wound; an effect not seen previously.

Statistical analyses on the neuronal data of the animals permitted only 1 week for recovery revealed significantly increased scores of 2+ or greater in the area of the lesion (versus sham operation + distal to the lesion) for IgG cyto-

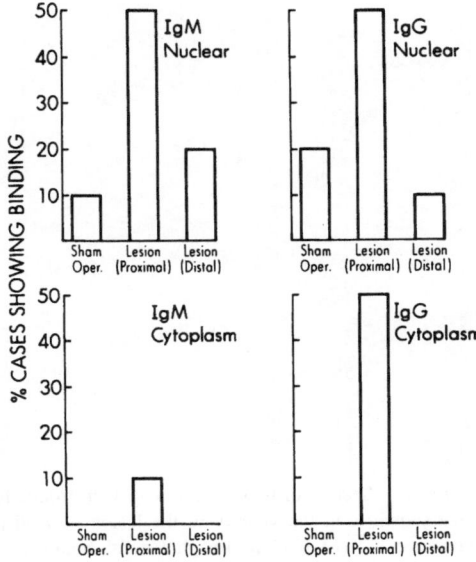

Figure 6. Percent of mice with frontal lobe transections or sham operations showing binding to each of four types of immunoglobulin 1 week after surgery. Data proximal and distal to the lesion are presented separately to show the locus specificity of the response. Each bar is based on the data from ten mice.

plasmic binding ($P < 0.005$). Borderline values of $P < 0.057$ were obtained for both IgM and IgG nuclear effects. No statistic could be applied to the dichotomized IgM cytoplasmic reaction since only one mouse showed positive binding of this type. However, even here the group mean was about twice that seen in mice given much longer recovery times.

These experiments thus show that there is a significant increase in antibodies binding to neurons following brain damage, and that this reaction can be observed in relatively young mice. The data also indicate that the neuronal binding seen after a brain lesion may involve at least two classes of immunoglobulin and nuclear as well as ctyoplasmic material. Glia also seem to be involved in the initial reaction to a wound of the brain (Fig. 7).

Nandy (1972) has claimed that in healthy C57 mice younger than 6 months of age, there is little evidence for the presence of circulating antibodies that have the potential to bind with neurons, although, as we have seen, an occasional 100-day-old control animal may show such binding to neuronal nuclei, if not to cytoplasm. The marked increase in frequency of both types of binding in the mice that were given brain lesions 1 week earlier would suggest that the damage resulted in the production of neuron-binding antibodies, and opened the

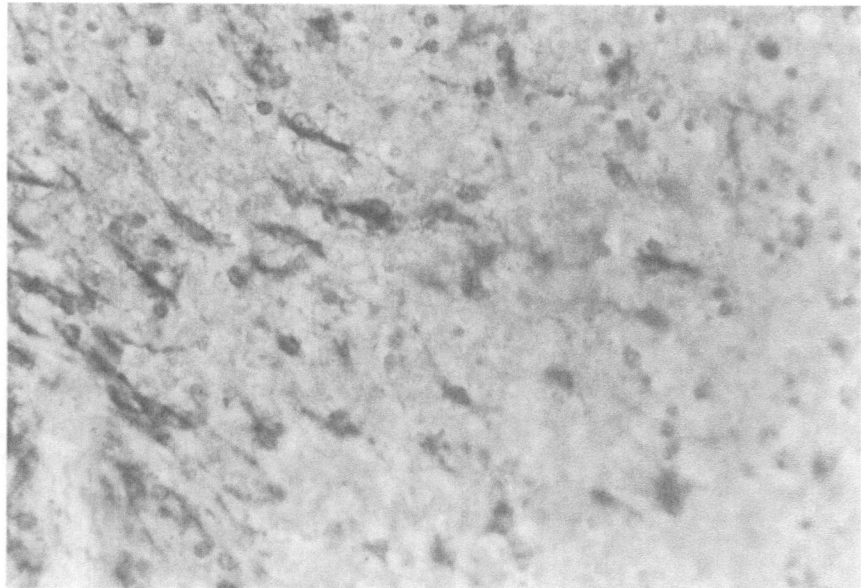

Figure 7. Section of brain of a mouse with frontal lobe transection sustained 1 week previously. Area shown is in the vicinity of the lesion (termed proximal in Fig. 6). The dark-staining cells indicative of IgM binding are glial cells. Staining is predominantly cytoplasmic. Magnification approximately 500×.

blood–brain barrier long enough for them to bind to cells in the region close to the lesion. Threatt *et al.* (1971) may have missed seeing this among his relatively young brain-damaged mice because they were sacrificed within a matter of hours after surgery—an inadequate amount of time for antibodies not already in the blood to be produced in sufficient amounts for binding to be detected at the level of the brain.

The observation that mice with lesions given long recovery periods did not show more binding than the control animals would suggest that the blood–brain barrier was able to repair itself. Changes in antibody production, and events that may have "neutralized" the binding response, however, may also have contributed to the drop in binding levels seen after the initial response. In any case, for reasons which are not yet clear, these brain-damaged mice showed even less binding to IgM nuclear antibodies than sham-operated mice in old age.

As mentioned above, an increase in neuronal IgM nuclear binding was found in the very old control mice. This finding supports the observation that some circulating antibodies with the potential to bind with brain tissue increase in frequency in old organisms including mice (Nandy, 1972), rats (Miller and Blumenthal, 1978), monkeys (Felsenfeld and Wolf, 1972), and humans (Felsenfeld and Wolf, 1972; Ingram *et al.*, 1974). However, in all but one of the above mentioned studies (Miller and Blumenthal, 1978), only "indirect" measurement procedures were utilized, and these showed only that the antibodies were present in the blood, not that they had already bound to the brain in sizeable numbers. On the basis of the "direct" immunoperoxidase procedures used here, as well as in other recent experiments from this laboratory (Baldinger and Blumenthal, 1982; Feden *et al.*, 1979; Miller and Blumenthal, 1978), it can now be concluded that at least some of these agents do in fact bind with neurons in increasing numbers later in life, but over a time course and in a manner that can vary greatly both across and within a species.

The binding seen among the sham-operated mice in the present study was almost exclusively to nuclear material [antinuclear antibodies (ANAs)]; cytoplasmic binding to IgG and IgM remained at very low levels, even late in life. Although Nandy and his co-workers (Threatt *et al.*, 1971; Nandy, 1972, 1975) have stated that the binding that they observed in the same strain of mouse later in life involved the nucleus, the cytoplasm, and the cell wall, the frequency of each type of binding was never analyzed or presented. Thus, the extent to which the present findings based on direct immunoperoxidase procedures parallel the earlier findings dealing with antibodies against neurons in serum is largely unknown. Serological examinations of humans and monkeys of different ages, however, have shown that nuclear binding antibodies that react with neurons increase in later life, and that agents in the blood that can bind to the cytoplasm of neurons may remain at low levels in these organisms even in old age (Felsenfeld and Wolf, 1972). While such observations are consistent with the findings reported in our own experiments, more recent studies utilizing direct assessment techniques have now associated increased cytoplasmic binding with aging in rats (Feden *et al.*, 1979; Miller and Blumenthal, 1978) and man (Baldinger and Blumenthal, 1982). These various findings may well reflect some species differences, but methodological and procedural variables, such as the criteria and assays used, and whether the brains were from subjects that had already been dead for some time, might also be contributing to these effects.

Both nuclear and cytoplasmic binding have been associated with neurological and psychiatric diseases in man, although it may be questioned whether the specific immunoglobulins involved are the same as those seen after an acute brain lesion or with aging. Still, it is interesting to note that cytoplasmic binding to neurons has been associated with Sydenham's chorea (Husby *et al.*, 1976) and carcinomatous neuropathy (Zeromski, 1970), and that nuclear binding to neurons has been observed in cases of myasthenia gravis (Martin *et al.*, 1974), senile dementia (D'Angelo and D'Angelo, 1975), and schizophrenia (Heath and Krupp, 1967). Walford (1969) has pointed out that increased nuclear binding has not been correlated with a specific disease in this strain of mouse. However, the *BDS* effect would suggest that under conditions of stress there may be a relationship between nuclear binding and disease, but because of small sample sizes further experimentation is necessary to clarify the nature of this association. In addition, the data suggest that neuron-binding may not necessarily be related to lifespan. Immediately after the lesion, for example, these mice showed intense nuclear and cytoplasmic binding, but as we have seen, the lesions had no effect on lifespan (Wozniak *et al.*, 1982). In addition, among the mice that had died from natural causes, those that showed binding at the time of death had lived just as long as those that did not.

The specificity of the nuclear binding that is observed at the level of the brain is rarely addressed in studies dealing with specific pathological conditions such as those mentioned above, or, for that matter, in most aging experiments. However, ANAs in the blood that increase in frequency with aging have been found to react with a number of other organs in studies not involving the brain (see Walford, 1967). A generalized response to nuclear-binding antibodies was confirmed in our aging experiments. In all mice that showed binding to the nuclei of neurons, there was also binding to at least one of the three other test organs; liver, kidney, and pancreas (Fig. 8). The ANA response seen with aging in C57 mice thus does not appear to be specific. This would imply that these particular antibodies might not even have originated as a reaction against brain tissue. Yet, why some subjects show such binding to the brain while others show none at all is not clear. Cytoplasmic binding, which was only seen here for a short period of time following the acute brain lesion, may be more specific, but this was not tested in these experiments. In rats, Miller and Blumenthal (1978) have demonstrated that antibodies against the cytoplasm of neurons show little cross-reactivity with other organs, except the thymus and perhaps the spleen.

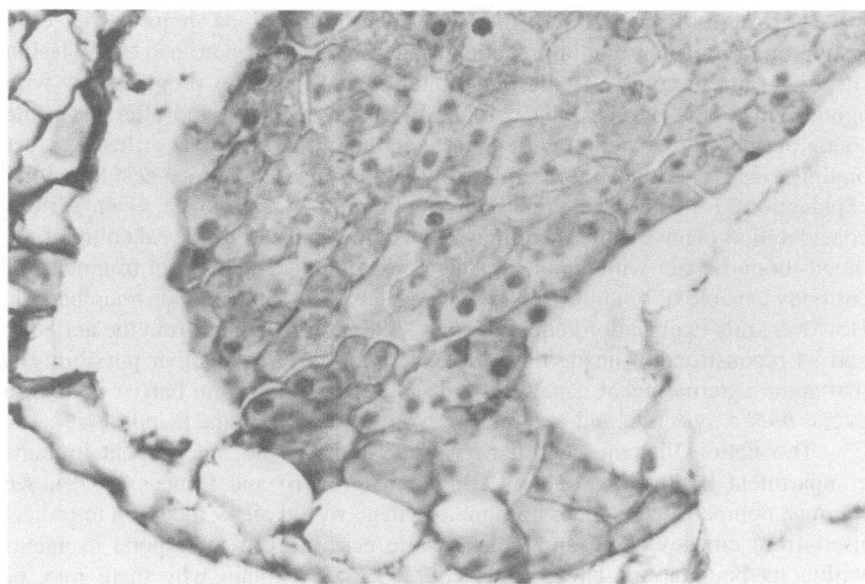

Figure 8. Section of pancreas of a mouse with positive nuclear IgM binding of neurons. Pancreatic acinar cells show similar nuclear binding. Magnification approximately 500×.

The functional significance of the binding which follows a brain injury is not well understood. The data collected here suggest that this response may have some adaptive value in that the antibodies seemed to be participating in the removal of dead or injured cells. The binding clearly was heaviest in the necrotic region close to the cut. Some seemingly healthy cells in the general vicinity of the lesion also showed some binding, but it cannot be stated with confidence that these cells were not also affected by the trauma (e.g., diaschisis, disrupted blood supply; Finger and Stein, 1982), and whether they would have remained healthy and viable under these conditions is an open question. Conceivably, some of the binding seen with aging could in part have a similar function. Still, whether the binding originated specifically to remove dead or injured cells can be questioned. In particular, age-related binding in some species seems to be taking place well after the organism has stopped contributing significantly to the gene pool. This, along with a low probability of survival from any type of a penetrating head wound, must be considered when the origins and functional significance of neuronal binding are put into evolutionary perspective.

Finally, the present data can be related to the popular notion that an increase in binding to neurons as a function of age represents a general weakening of the blood–brain barrier (Threatt *et al.*, 1971). It was observed that when a cut was made across the frontal cortex, the local break in the blood–brain barrier only resulted in a transient binding reaction, and one basically confined to the lesion site—i.e., not the type of widespread binding that increases progressively with aging. This, plus the fact that cytoplasmic binding was found after the acute lesion but not in association with aging, and the observation that diffuse nuclear binding was sometimes seen in relatively young mice, would suggest that other explanations for age-related increases in binding must also be given serious consideration. Thus, while there may indeed be some sort of a weakening of the blood–brain barrier with aging, it is also possible that monovalent fragments of antibody capable of binding with antigen can cross an intact blood–brain barrier. However, this explanation requires a dissociation of fragment from the antibody and its reconstruction inside the central nervous system. Another possibility is that some external agent capable of crossing the blood–brain barrier is able to evoke *both* a systemic and a central nervous system immune response.

The notion that the central nervous system has an independent immune compartment has been raised by Blumenthal (1976) and Prineas (1979). An immune compartment and an immune privilege would allow the brain to protect itself from cross-reacting antibodies, while permitting it to respond to agents within its boundaries. These ideas could help to explain why there may be significant levels of some antibodies in the blood with the potential to react with brain tissue, but more limited binding to the brain itself.

5. CONCLUSIONS

The purpose of the present series of experiments was to look at the effects of brain damage on lifespan, morbidity, and neuron-binding antibodies, three dependent variables rarely examined in the clinical or experimental literatures dealing with brain damage or aging. Using the frontal cortex of the mouse as a model, it was observed that these knife cuts did not affect lifespan or the diseases that the animals contacted and eventually died from many months or some years later. An antibody response to the brain, however, was noted after the blood–brain barrier was damaged. However, within a matter of months the number of mice with lesions showing such binding returned to control levels, and late in life these mice even appeared to show less binding than sham-operated animals to antibodies that are associated with aging and a hypothesized "weakening" of the blood–brain barrier.

The full significance of the antibody phenomenon is not yet clear, although the data collected in these experiments seem to suggest that the presence of such binding is not associated with a shorter lifespan. In any event, the blood–brain barrier probably plays an important role here, and in this regard it is important to remember that cutting the cerebral cortex with a scalpel blade may not be directly comparable to the type of destruction that would result from certain chronic diseases of the central nervous system or impairment of its blood supply.

The findings reported here obviously raise at least as many questions as they answer. For example, would lesions of the hypothalamus, hippocampus, or midbrain central gray have similar effects? Would damage to other cortical areas also result in a transient antibody reaction without affecting morbidity and longevity? Would the reactions be the same if inflicted in subjects much younger or older than those used here? And, in species showing cytoplasmic binding as a function of age, would such binding be associated with morbidity? These are some of the questions about which we can only speculate at the present time. Nevertheless, they are important if for no other reason than to remind us how little we still know about aging and age-related pathology.

From a different but related perspective, the questions raised by these experiments also reflect the fact that much still remains to be learned about the consequences of brain damage, and especially those phenomena and events that do not have obvious behavioral correlates, or whose correlates may be subtle and manifested only long after the initial effects of the injury have subsided. It would be naive to believe that the changes that accompany brain damage end or stabilize a few weeks or months later when a patient is released from a hospital or completes a program of therapy, or when a group of laboratory animals finishes a battery of behavioral tests. Looking beyond the immediate or overt effects of a lesion may represent a formidable task, but the task is important and it is one that can be rewarding and enlightening. Viewing the consequences of a brain lesion within a lifespan framework is essential for understanding such things as normal aging, evolution, adaptation, and survival. A thorough understanding of more than just the immediate, overt effects of brain damage is also necessary for choosing the best diagnostic treatment and therapeutic interventions in the health-care professions—interventions that we hope will be effective and yet ones that will afford minimal risk to the patient.

ACKNOWLEDGMENTS.   This research was supported by NINCDS Grant 11002 to the first author. The assistance of Russell Poland (corticosterone assays), Douglas Young (immunoglobulin assays), and Edward Spitznagel, Robert Goldstein, and Andrew Finn (statistical analyses) is gratefully acknowledged.

REFERENCES

Baldinger, A., and Blumenthal, H. T., 1982, Neuroimmunology of the aging brain, in: *Geriatrics*, Volume 1 (D. Platt, ed.), Springer-Verlag, Berlin, pp. 283–299.
Blumenthal, H. T., 1976, Immunological aspects of the aging brain, in: *Neurobiology of Aging* (R. D. Terry and S. Gershon, eds.), Raven Press, New York, pp. 313–334.

Blumenthal, H., Young, D., Wozniak, D., and Finger S., 1984, Effects of age, brain damage and stress on antibody binding to the brain, *J. Gerontol.* (in press).

Brody, H., 1955, Organization of the cerebral cortex. III. A study of aging in the human cerebral cortex, *J. Comp. Neurol.* **102**:511–556.

D'Angelo, C., and D'Angelo, D. B., 1975, Possibility of an autoimmune pathogenesis of some histologic alterations, characteristics in presenile and senile dementia, in: *Neuropathology,* Volume 2 (S. Kornyey, S. Tariska, and G. Gosztonyi, eds.), Excerpta Medica, Amsterdam, pp. 131–134.

Denkla, W. D., 1977, Systems analysis of possible mechanisms of mammalian aging, *Mech. Ageing Dev.* **6**:143–152.

Dilman, V. M., 1976, The hypothalamic control of aging and age associated pathology: The elevation mechanism of aging, in: *Hypothalamus, Pituitary and Aging* (A. V. Everitt and J. A. Burgess, eds.), Charles C. Thomas, Springfield, Illinois, pp. 634–667.

Dilman, V. M., 1979, Hypothalamic mechanisms of aging and of specific age pathology. V. A model for the mechanism of human-specific age pathology and natural death, *Exp. Gerontol.* **14**:287–300.

Feden, G., Baldinger, A., Miller-Soule, D., and Blumenthal, H. T., 1979, An in vivo and in vitro study of an aging-related neuron cytoplasmic binding antibody in male Fischer rats, *J. Gerontol.* **34**:651–660.

Felsenfeld, G., and Wolf, R. F., 1972, Relationship of age and serum immunoglobulins to autoantibodies against brain constituents in primates. I. A study in apparently healthy man, Macaca mullata and Erythrocebus pates, *J. Med. Primatol.* **1**:287–296.

Fienberg, S. E., 1979, *The Analysis of Cross-Classified Categorical Data,* MIT Press, Cambridge.

Finch, C. E., 1976, The regulation of physiological changes during mammalian aging, *Q. Rev. Biol.* **51**:49–83.

Finger, S., and Stein, D. G. (eds.), 1982, *Brain Damage and Recovery: Research and Clinical Perspectives,* Academic Press, New York.

Frolkis, V. V., 1976, The hypothalamic mechanisms of aging, in: *Hypothalamus, Pituitary and Aging* (A. V. Everitt and J. A. Burgess, eds.), Charles C. Thomas, Springfield, Illinois, pp. 614–633.

Frolkis, V. V., Bezrukov, V. V., Duplenko, Yu. K., and Genis, E. D., 1972, The hypothalamus in aging, *Exp. Gerontol.* **7**:169–184.

Glick, S. D., Nakamura, R. K., and Jarvik, M. E., 1971, Recovery of function following frontal brain damage in mice, *J. Comp. Physiol. Psychol.* **76**:454–459.

Groen, J. J., 1959, General physiology of aging, *Geriatrics* **14**:318–331.

Heath, R. G., and Krupp, I. M., 1967, The biological basis of schizophrenia. An autoimmune concept, in: *Molecular Basis of Some Aspects of Mental Activity* (O. Walaas, ed.), Academic Press, New York, pp. 313–344.

Husby, C., Riji, I. V. E., Zabriskie, J. B., Abdin, K. M., and Williams, R. C., 1976, Antibodies reacting with cytoplasm of subthalamic and caudate nuclei neurons in chorea and acute rheumatic fever, *J. Exp. Med.* **144**:1095–1110.

Ingram, C. R., Phegan, K. J., and Blumenthal, H. T., 1974, Significance of an aging-linked neuron binding gamma globulin fraction of human sera, *J. Gerontol.* **29**:20–27.

Kolyaskina, G. I., and Kushner, S. G., 1969, Principles governing appearance of anti-brain antibodies in serum of schizophrenics, *Neuropathol. Psychiatr.* **69**:1679–1682.

Liu, R. K., and Walford, R. L., 1972, The effect of lowered body temperature on life-span and immune and non-immune process, *Gerontologia* **18**:363–388.

Martin, I., Herr, J. C., Wanamaker, B. A. W., and Kornguth, S., 1974, Demonstration of specific antineuronal nuclear antibody in sera of patients with myasthenia gravis. Indirect and direct immunoflourescence, *Neurology* **24**:680–684.

Miller, D. T., and Blumenthal, H. T., 1978, Neuron-thymic lymphocyte binding of serum IgG of 90- and 500-day-old female Wistar albino rats, *J. Gerontol.* **33**:129–136.

Motyka, A., and Jezkova, A., 1975, Autoantibodies and brain ischemia tomography, *Cas. Lek. Cesk.* **114**:1455–1457.

Nandy, K., 1972, Brain-reactive antibodies in mouse serum as a function of age, *J. Gerontol.* **27**:173–177.

Nandy, K., 1975, Significance of brain reactive antibodies in serum of old mice, *J. Gerontol.* **30**:412–416.

Nandy, K., 1977, Immune reactions in aging brain and senile dementia, in: *The Aging Brain and Senile Dementia* (K. Nandy and I. Sherwin, eds.), Plenum Press, New York, pp. 181–196.

Ordy, J. M., Samorajski, T., Zeman, W., and Curtis, H. J., 1967, Interaction effects of environmental stress and deuteron irradiation of the brain on mortality and longevity in C57BL-10 mice, *Proc. Soc. Exp. Biol. Med.* **126**:184–190.

Poland, R. E., Weichsel, M. E., Jr., and Rubin, R. T., 1981, Neonatal dexamethasone administration. I. Temporary delay of development of the circadian serum corticosterone rhythm in rats, *Endocrinology* **108**:1049–1054.

Prineas, J. W., 1979, Multiple sclerosis: Presence of lymphatic capillaries and lymphoid tissue in brain and spinal cord, *Science* **203**:1123–1125.

Rolsten, C., Ordy, J. M., and Samorajski, T., 1973, Age and sex differences in mortality after proton irradiation of the brain during maturity and aging in C57BL/10 mice, *J. Gerontol.* **28**:460–465.

Rowlatt, C., Chesterman, F. C., and Sheriff, M. U., 1976, Lifespan, age changes and tumor incidence in an aging C57BL mouse colony, *Lab. Anim.* **10**:419–442.

Sacher, G. S., 1977, Life table modification and life prolongation, in: *Handbook of the Biology of Aging* (C. E. Finch and L. Hayflick, eds.), Van Nostrand Reinhold, New York, pp. 582–638.

Samorajski, T., Ordy, J. M., Zeman, W., and Curtis, H. J., 1970, Brain irradiation and aging, *Interdiscip. Top. Gerontol.* **7**:72–86.

Savenko, S. N., and Polienko, E. M., 1975, Autoimmune factors in cerebral circulatory disorders, *Zh. Neuropatol. Psykhiatr.* **75**:3–7.

Sklar, L. S., and Anisman, H., 1979, Stress and coping factors influence tumor growth, *Science* **205**:513–515.

Stein, D. G., and Firl, A. S., 1976, Brain damage and reorganization of function in old age, *Exp. Neurol.* **52**:157–167.

Threatt, J., Nandy, K., and Fritz, R., 1971, Brain-reactive antibodies in serum of old mice demonstrated by immunofluorescence, *J. Gerontol.* **26**:316–323.

Timiras, P. S., 1978, Biological perspectives on aging, *Am. Sci.* **66**:605–613.

Vogel, F. S., 1969, The brain and time, in: *Behavior and Adaptation in Later Life* (E. W. Busse and E. Pfeiffer, eds.), Little, Brown, Boston, pp. 251–262.

Walford, R. L., 1967, *The General Immunology of Aging. Advances in Gerontological Research*, Volume 2 (B. L.Strehler, ed.), Academic Press, New York, pp. 159–204.

Walford, R. L., 1969, *The Immunologic Theory of Aging*, Williams and Wilkins, Baltimore.

Wozniak, D. F., and Goldstein, R., 1980, Effect of deprivation duration and prefeeding on gastric stress erosions in the rat, *Physiol. Behav.* **24**:231–235.

Wozniak, D., Finger, S., Blumenthal, H., and Poland, R., 1982, Brain damage, stress and lifespan: An experimental study, *J. Gerontol.* **37**:161–168.

Zeromski, J., 1970, Immunological findings in sensory carcinomatous neuropathy. Application of peroxidase labelled antibody, *Clin. Exp. Immunol.* **6**:633–637.

2

Aging and Hypothalamic Regulation of Metabolic, Autonomic, and Endocrine Function

C. ROBERT ALMLI

1. INTRODUCTION

The aging process is a complex and mysterious phenomenon that appears to affect all mammals as they evolve through senescence and ultimately death. During the aging process, widespread and heterogeneous changes may occur in a variety of functions including cognitive, mental, sensory, motor, metabolic, autonomic, and/or endocrine function. In spite of the fact that a variety of functions may be affected during the aging process, there are tremendous individual and species differences with respect to the pattern of functional changes displayed during aging, and furthermore the functional changes associated with aging may be displayed in a heterogeneous fashion within individuals, i.e., the whole body does not age at the same rate. The widespread and diverse nature of the functional changes that may be associated with aging underscores the complexity of the aging process. This complexity has served to reinforce the notion that aging, and the rate of aging, are products of some genetic program–environment interaction.

The complexity and diversity of the aging process probably account for the fact that there is no universally accepted definition of aging. However, many of the proposed definitions of aging include one or more of the following: progressive deterioration of the organism, involution of the organism over time, in-

C. ROBERT ALMLI • Departments of Preventive Medicine, Anatomy and Neurobiology, and Psychology, Programs in Occupational Therapy and Neural Sciences, Washington University School of Medicine, St. Louis, Missouri 63110.

creased wear and tear of the organism, or increased probability of death. Each of these definitional components seems to indicate that aging represents a progressive decrease in function or adaptability of the organism, and this loss would eventually be associated with aging-related pathology and/or death. While death due to "old age" may occur, death in aging is often associated with aging-related pathology such as hypertension, diabetes, and infection (Everitt, 1976a).

Research on the regulation of aging and aging-related pathology has been carried out for many years, and much of this research has been focused on the role of the nervous system in the regulation of the aging process. Because of the complexity and diversity of the functions that may be altered during aging (e.g., cognitive, mental, sensory, motor, metabolic, autonomic, endocrine), it seems obvious that the entire nervous system is probably involved in the regulation or manifestation of the aging process. The issue to be addressed in the present chapter is an evaluation of the role played by the hypothalamus in the regulation of aging-related changes in metabolic, autonomic, and endocrine function.

The hypothalamus has received attention in the aging literature for at least three reasons. First, it is well known that the hypothalamus is involved in the regulation of metabolic, autonomic, and endocrine function. Second, metabolic, autonomic, and endocrine functions are often changed or disrupted during the aging process. And third, there is a close resemblance between aging-related changes in metabolic, autonomic, and endocrine function and the clinical manifestation of hypothalamic disease.

In the initial sections of this chapter, a selection of various theories of aging are presented, and this is followed by a survey of aging-related changes in metabolic, autonomic, and endocrine function. In later sections, the structure and functions of the hypothalamus are summarized, and then three theories of hypothalamic regulation of aging are presented. In the final sections of this chapter, the effects of experimental damage to the hypothalamus (lateral hypothalamus and ventromedial hypothalamus) are surveyed, and the metabolic, autonomic, and endocrine effects of hypothalamic damage are compared to the metabolic, autonomic, and endocrine changes associated with the aging process and aging-related pathology. In this comparison it will be assumed that experimental hypothalamic damage produces dysfunction, and this hypothalamic dysfunction to some extent approximates the hypothalamic dysfunction that is the basis of the various theories of hypothalamic regulation of aging. A second assumption that will be made is that the behavioral and physiological changes associated with hypothalamic damage and those changes associated with aging and aging-related pathology are consistent and reliable effects (i.e., species and individual differences are ignored at this time). Obviously these assumptions can be questioned. However, it is necessary to make these assumptions for this initial analysis of the hypothalamic theories of aging.

1.1. Theories of Aging

The theories of aging that have been proposed over the years are generally broad based and diverse, and have been difficult to evaluate scientifically. However, many theorists agree that aging and longevity are at least partially genetically determined (Comfort, 1964), and that environmental factors interact with genetic programs of aging and thus affect the rate of the aging process.

The diversity of the various theories of aging is exemplified by their wide-ranging differences in emphasis. For example, there are theories that emphasize genetic programs (Comfort, 1964), random mutations in the DNA of somatic cells, mutated cells that stimulate immunological reactions against the body's own tissues, rate of living, and reactivity to environmental stressors. These representative theories of aging are not mutually exclusive, and it is becoming more apparent that the primary control of the aging process most likely resides within the genetic make-up of the individual, and that the rate of aging is a product of some genetic program–environment interaction (Everitt, 1976c). The genetic program–environment interaction for the aging process is thought by some to be at least partially mediated through the hypothalamus. Selected hypothalamic theories of aging are presented later in this chapter.

1.2. Aging-Related Changes in Behavioral and Physiological Function

The aging process is associated with a wide variety of changes in metabolic, autonomic, and endocrine function, and many of the changes are detrimental to organismic adaptability and, therefore viability. For example, the aging process is typically associated with decreases in maximal breathing capacity, cardiac output, and a decreased ability of the individual to adapt to environmental stressors. There may also be decreases in sensory, motor, and cognitive abilities (Potvin et al., 1980).

Some representative morphological changes frequently associated with aging are decreased weights of the kidney, thyroid gland, adrenal glands, testes, ovaries, uterus, liver, pancreas, and skeletal muscle; decreased growth rate (McCay et al., 1935); increased body fat content (Young et al., 1963); and increased heart size. Hormonal alterations frequently associated with aging are decreased secretion of the anterior pituitary gland, decreased secretion of growth hormone (Finkelstein et al., 1972), increased secretion of prolactin, decreased secretion of antidiuretic hormone, and decreased thyroid activity (Everitt, 1976b). Other physiological changes typically associated with aging are decreased glomerular filtration rate, increased protein in urine, increased systolic blood pressure, altered serum cholesterol levels (Keys, 1952), increased lipids in fat cells, and deterioration in glucose tolerance. These various functional alterations are found to be associated with aging in a variety of species including man.

Further, research conducted with humans and other mammals has revealed a variety of changes in the morphology and chemistry of the aging brain. It has been suggested that these brain changes may underlie some of the functional changes associated with aging. Some of the more frequently reported aging-related (eugeric or pathogenic) changes in the human and animal brain are a decrease in brain weight (Pearl, 1905), shrinkage of cortical gyri, and deepening of cortical sulci. With age, there is often a decrease in brain water, RNA, protein, and amino acid concentration or turnover rate (Ordy and Kaack, 1975), with a reduction in cell nucleus size. Further, many brain regions may show increased accumulation of cellular lipofuscin pigments with age (Bourne, 1973), nerve cell loss (Brody, 1973), and increased numbers of glial cells. In addition, neurons may show dendritic swelling, distortion, shortening, and degeneration during aging (Scheibel and Scheibel, 1977). Aging is also often associated with loss of dendritic spines (Scheibel and Scheibel, 1977). Further, axonal dystrophy, swelling, and degeneration have been reported with aging, as well as terminal bouton vesicular losses and reductions of axosomatic synapses (Berlin and Wallace, 1976). Aging may also be associated with decreased intercellular contacts, reduced myelin lipids, and decreased neurotransmitter synthesis and/or increased neurotransmitter catabolism (Miller et al., 1976).

Thus, aging in a variety of mammals, including man, is associated with widespread changes in structure and function. Although the patterns of aging-related changes in structure and function are becoming more clear, it is also apparent that there is tremendous variability between and within individuals and species with respect to the pattern and timing of aging-related changes. These individual and species differences most likely represent the product of the genetic–environment interaction influencing the pattern and rate of aging. It is a given that chronological age does not necessarily equal biological age, and this inequality has contributed to the tremendous variability found in much of the aging research.

From the above presentation of the widespread structural and functional changes associated with aging, it is obvious that the aging process is a truly complex phenomenon. Further, many of the structural and functional changes of the aging individual may be considered pathological or at least prepathological, i.e., the changes serve to reduce the organism's adaptability and chances of survival. When aging is viewed in this manner, the question of whether aging is a health or disease process becomes apparent. This issue will not be addressed in this chapter. However, it is becoming increasingly clear that while the aging process may be facilitated or delayed to some extent, aging is often associated with some eventual form of pathology (e.g., cancer, cardiovascular disease, renal disease) followed by death of the organism.

Because "aging" is a fact of life (Comfort, 1964), much research has been focused upon describing the structural and functional changes that are associated with the aging process as has been presented above. The ultimate target of such research is to elucidate the etiology or etiologies of the aging process and aging-related pathology. In aging, there consistently appears to be a decline in the individual's ability to adapt to changes within the internal and external environments. Because of this aging decline and because many of the aging-related pathologies resemble the clinical manifestation of hypothalamic disease, the hypothalamus has received much attention with respect to its possible role in aging and aging-related pathology.

2. STRUCTURE AND FUNCTION OF THE HYPOTHALAMUS

It is well known that the hypothalamus is an anatomically, biochemically, and physiologically complex region of the brain, and that this heterogeneous neural region has vast interconnections with the remainder of the nervous system (see Morgane and Panksepp, 1979, 1980a, 1981). The hypothalamus, like most other brain regions (e.g., the pons), participates in the regulation of a variety of behavioral and physiological functions. While the hypothalamus is obviously important in the regulation of many behavioral and physiological functions, it does not appear to be an absolute controller of any specific function. The hypothalamus is hardly a "brain within the brain," in spite of the fact that the hypothalamus may play a significant role in the integration of environmental influences and bodily states related to homeostatic function and the adaptation process.

The hypothalamus has interconnections with the limbic system, and some researchers consider the hypothalamus to be a part of the limbic system. The limbic system is well known to be involved in emotional reactivity and stress responsivity (Papez, 1937). Further, the hypothalamus has interconnections with the autonomic nervous system (Ban, 1975) and with the endocrine system via various neural, hormonal, and releasing factor influences (Knigge et al., 1980). Thus, the hypothalamus appears to occupy a relatively strategic position to play a role in the integration or regulation of behavior and physiology with respect to alterations within the external and internal environments of the individual.

Considerable laboratory research with a variety of animals has revealed that the hypothalamus plays a role in the regulation of metabolic, autonomic, and endocrine function. The hypothalamus has been shown to play a role in the regulation of feeding behavior and metabolism (Oomura, 1976), drinking behavior and body water, blood pressure (Ranson and Magoun, 1939), the electrocardiogram, epinephrine and renin secretion, the gastrointestinal system, and reproductive behaviors and hormones (Pfaff, 1981). For a detailed discussion of hypothalamic anatomy and physiology, the reader is referred to Morgane and Panksepp (1979, 1980a,b, 1981).

Thus, the hypothalamus appears to be involved in the regulation of those metabolic, autonomic, and endocrine processes that are often altered during aging and aging-related pathology. Because of this suggestive parallel, it is not surprising that the hypothalamus has been considered to be invovled in the regulation of aging and an important mediator of aging-related pathology. In the next section selected theories of hypothalamic regulation of the aging process are presented.

3. HYPOTHALAMIC THEORIES OF AGING

Everitt (1976a) and many others have suggested that the interaction between genetic programs and environmental factors relating to the process of aging may be mediated through the neuroendocrine system, or more specifically, through the hypothalamic–pituitary–peripheral endocrine system axis. The pituitary has long been considered to be a primary factor in the regulation of the aging process due to its role as a regulator of peripheral endocrine glands, as well as its role in the regulation of growth, reproduction, and metabolism. The hypothalamus, as a regulator of pituitary and peripheral endocrine gland function, is suggested to occupy a pivotal position within this axis. The hypothalamus, via hormonal feedback loops and through its many and complex neural connections, is a recipient of information related to the status of the organism's internal and external environments (Fisher and Almli, 1979). The hypothalamus, therefore, appears to be in a strategic position to play a major role in the integration of environmental stimulation and bodily states related to the aging process, not only for the neuroendocrine system axis, but for the autonomic and somatic systems as well.

The role of the hypothalamus as a regulator of the aging process has been highlighted in numerous hypothalamic theories of aging. These hypothalamic theories are the result of the realization that there are many similarities between a number of aging characteristics and the clinical manifestation of hypothalamic disease. It has been suggested that age-related changes occur within the hypothalamus (e.g., changes in neurotransmitters, enzymes, and metabolites) and that these changes are important for aging and cell death (Samorajski, 1977). More specifically, there have been reports of abnormal neurofibrillary changes within the hypothalamus upon postmortum examination of amyotrophic lateral sclerosis and parkinsonian patients (Hirano et al., 1967). In the laboratory, Gutstein et al. (1978) electrically stimulated the lateral hypothalamus of rats and found changes reducing the lumen size of various arteries (e.g., plaques, thickening). These changes closely resembled the changes in arteries seen in human atherosclerosis. Finally, Korneva (1976) has suggested that hypothalamic lesions in humans may result in hyperreactivity to immunogens thereby producing allergy or hyporeactivity to immunogens and decreased resistance to disease. Thus, there is considerable experimental and clinical data which suggest a role for the hypothalamus in the aging process and aging-related pathology.

With respect to the aging process and aging-related pathology, there has been some interesting research on hypothalamic involvement in the regulation of the immune system. Experimental lesions of the hypothalamus of a wide variety of animals have been shown to result in changes in immune system responses, e.g., decreased antibody production, protection against lethal anaphylaxis (Paunovic *et al.*, 1976), and there appears to be some degree of regionalization within the hypothalamus for lesion loci that are effective for altering the function of the immune system. Also, the hypothalamic lesion effects upon the immune system have been shown to persist for long periods of time when hypothalamic lesions are sustained by infant animals (Paunovic *et al.*, 1976). It has been suggested that these lesion-induced immune system changes may be mediated through the autonomic nervous system and the neuroendocrine system (Stein *et al.*, 1976). However, it must also be noted that the effects of hypothalamic lesions on the immune system are not always positive and many negative reports have been made. Additional research is required to determine if the hypothalamus (or even the central nervous system for that matter) is a regulator, or merely a recipient of, immunological changes during the aging process.

In the next sections of this chapter are presented three of the major hypothalamic theories of aging and aging-related pathology. The basis of each of these theories is that some genetically programmed intrinsic and/or pathological changes within the hypothalamus may be causally related to the mechanisms of functional and metabolic changes associated with the aging process. In aging, Dilman (1976, 1979) proposes that hypothalamic activity is increased, Groen (1959) proposes that hypothalamic functional activity is reduced, and Frolkis (1976) and co-workers (1972) propose that disproportionate age changes within the hypothalamus result in hypothalamic disregulation of the individual. Although these theories appear to be contradictory, they may actually differ only in points of emphasis.

3.1. Dilman's Hypothalamic Elevation Theory

Dilman (1976, 1979) has proposed that an age-associated elevation of the threshold of sensitivity of the hypothalamic–pituitary complex to homeostatic stimuli plays a major role in the mechanisms of development, aging, and age-pathology formation. According to this theory, the hypothalamic "set point" of sensitivity to the three main homeostatic systems (energy, reproduction, and adaptation) becomes elevated during aging such that the hypothalamus becomes less sensitive to feedback suppression by glucose, estrogens, and adrenocortical steroids. This results in an elevation of hypothalamic thresholds which leads to an imbalance of the internal environment. In the theory, imbalance of the internal environment leads to homeostatic failure and pathology characteristic of old age, such as relative obesity, adult-onset diabetes mellitus, atherosclerosis, hypertensive disease, mental depression, decreased resistance to infection, and cancer.

This theory further proposes that the elevated hypothalamic thresholds to homeostatic stimuli result in a compensatory alteration of function by peripheral organs and glands which leads to elevation of serum cholesterol level, decreased glucose tolerance, increased blood insulin levels, increased cortisol levels, and climacteric. These aging changes are also associated with decreased growth hormone levels and an intensification of free fatty acid (FFA) utilization. The above changes associated with aging only partially characterize the complex hormonal-metabolic disturbances inherent in aging according to Dilman (1976, 1979).

Dilman (1976, 1979) thus proposes that aging is a genetically preprogramed process of derangement of homeostasis whereby hypothalamic thresholds to homeostatic stimuli are elevated. According to this theory, environmental factors can also interact with the genetic program and thus influence hypothalamic activity. For example, stress increases hypothalamic activity and facilitates the aging process, while reduced food intake decreases hypothalamic activity and attenuates the aging process (Dilman, 1976, 1979). The aging process and the diseases of aging, according to this theory, are the consequences of the organism's ability to adapt and the adaptation process.

3.2. Groen's Hypothalamic Theory

Groen (1959) has proposed a slightly different theory relating the hypothalamus to the aging process. Groen suggests that senile involution or senile obesity are a result of hypothalamic-mediated changes in appetite and hormone (gonadal and growth) function, i.e., a differential exhaustion of hypothalamic function. The hypothalamic changes may be induced by genetic–environment interaction, and they may represent an accumulation of sequelae of certain lesions (hypothalamic?) which do not give rise to obvious clinical disturbances until aging is well advanced. The aging process sets in at the time of arrest of cell division, and the process is cumulative throughout the life span of the organism.

Groen (1959) further proposes that aging-related changes within the hypothalamus may become manifest in two different patterns of aging. One type of change within the hypothalamus may result in decreased appetite, decreased gonadal function, decreased pituitary function, and so on. This pattern of hypothalamic change would result in hypometabolism, and thus, senile atrophy or involution. Hypothalamic regulation of senile involution is thus mediated via a decreased metabolic rate, and the effects are somewhat similar to the effects upon metabolism and longevity of food restriction experiments (McCay et al.,

1935). Senile involution may therefore be biologically adaptive in that some forms of aging-related pathology are delayed and longer life is promoted. With a different pattern of age-related hypothalamic change, the opposite of involution may result. A decreased lifespan would be associated with hypothalamic changes promoting hypermetabolism, which would result in overeating and obesity, and early onset of some forms of aging-related pathology.

3.3. Frolkis's Hypothalamic Disregulation Theory

The third hypothalamic theory of aging is that of Frolkis (1972, 1976). Frolkis also proposes that hypothalamic changes determine the important mechanisms of aging such as shifts in metabolism and adaptive capacity. Specifically, functional disregulation of various regions of the hypothalamus is thought to create the conditions of aging-related pathology such as arterial hypertension, atherosclerosis, myocardial infarction, diabetes, and pathological climacteric.

Frolkis (1976) based this theory on experimental (rabbits, rats) research showing that aging is associated with hypothalamic disregulation as expressed by diverse changes of the functions of separate structures within the hypothalamus. Hypothalamic changes associated with aging are a reduction of neurosecretory activity of the hypothalamic–hypophyseal system, and age-related differentials in the neurosecretory process in response to reflex (painful electrical stimulation) and humoral (adrenaline) stimulation. In addition, aging is associated with an increase in the electroexcitability of the anterior and posterior hypothalamus, with a concurrent decrease in the electroexcitability of the lateral hypothalamus. Further, there is an age-related increase in the sensitivity of the hypothalamus to the direct effect of adrenaline and acetylcholine, and finally, there are age-related changes in the effect of hypothalamic stimulation on blood circulation, hypertensive reactions, and cardiac output (Frolkis, 1976).

These aging-related changes within the hypothalamus are proposed to influence the quality of hypothalamic regulation of metabolism and other functions (Frolkis, 1976). Thus according to Frolkis, the aging process is based upon disregulation of the hypothalamus which results in disturbances in metabolism and function, decreased adaptive capacity, and age-related pathology.

The three theories presented above are similar in that each theory proposes that intrinsic (genetically programmed, exhaustive from use, accumulated lesions) changes within the hypothalamus form the basis of hypothalamic regulation of some of the characteristics of aging and aging-related pathology. Each of the theories also emphasizes the importance of changes in the hypothalamic regulations of metabolism during the aging process, and all three theories are similar in that they are oriented toward aging-related pathology. The three theories primarily differ in the "specific" types of changes proposed to occur within the hypothalamus, i.e., threshold changes, pathological or exhaustive changes, or divergent functional changes in different parts of the hypothalamus. However, these "specific" hypothalamic changes do not appear to be mutually exclusive, and it is conceivable that the changes may have a common source.

The major problem with each of the hypothalamic theories is they lack a data base for the proposed changes in hypothalamic function. All of the theories rely heavily on a proposed interaction between environmental factors and genetic programs to account for the aging-related changes in hypothalamic regulation of metabolic, autonomic, and endocrine function. This makes it difficult to ascertain if aging-related hypothalamic changes are "cause" or "effect," i.e., whether aging changes within the hypothalamus induce the metabolic, autonomic, and endocrine manifestations of aging, or whether hypothalamic aging-related changes are merely consequences of aging of other systems of the body.

4. EFFECTS OF HYPOTHALAMIC DAMAGE

Each of the theories presented above propose that the hypothalamus is a regulator of some functional changes associated with aging and aging-related pathology, and this influence on aging is related to the role of the hypothalamus as a regulator of autonomic, metabolic, and endocrine function. If the hypothalamus is a regulator of aging via genetically programed dysfunction, exhaustion, and/or accumulated lesions, then it might be hypothesized that experimental studies of hypothalamic damage should yield results that would indicate a facilitation or retardation of some characteristics of the aging process and/or aging-related pathology. In the next sections of this chapter, the behavioral and physiological effects of damage inflicted into the hypothalamus of a variety of species are outlined. The two regions of the hypothalamus that are presented are the lateral hypothalamic area and the ventromedial hypothalamic area. These hypothalamic regions were chosen because of the relatively large volume of literature available on the behavioral and physiological effects of experimental damage to these areas. However, the reader is reminded that the "lateral hypothalamus" and "ventromedial hypothalamus" are complex neural regions consisting of intrinsic neurons and numerous fiber projection systems. This means that experimental lesions of these areas produce diffuse neural damage which goes beyond the targeted lesion locus. Thus, the behavioral and physiological changes that occur following damage to these hypothalamic regions may be related to disruption of different neural systems (intrahypothalamic and extrahypothalamic) that are affected, directly or indirectly, by such gross damage to the hypothalamus. The inclusion of other hypothalamic regions in the present discussion would have been appropriate. They are excluded only because of space limits.

If the hypothalamus plays a role in the regulation of aging and aging-related pathology because of some form of hypothalamic dysfunction, then experimental damage to the hypothalamus might be expected to produce an approximation of some of the behavioral and physiological changes associated with aging and aging-related pathology. In the final sections of this chapter the behavioral and physiological effects of lateral hypothalamic and ventromedial hypothalamic damage sustained by a variety of species are surveyed. This is followed by a comparison of the behavioral and physiological effects of experimental hypothalamic damage and the metabolic, autonomic, and endocrine manifestations of aging and aging-related pathology. The various hypothalamic theories of aging would lead to the prediction of considerable correspondence between the effects of experimental hypothalamic damage and some of the characteristics of the aging process.

4.1. Lateral Hypothalamic Damage

Research has shown that damage inflicted into the lateral hypothalamic (LH) region of a variety of species (adults) results in considerable change in metabolic, autonomic, and endocrine function. Following LH damage, animals display disruption of ingestive (feeding and drinking) behaviors (Almli and Weiss, 1975) and reduction of body weight (Powley and Keesey, 1970). The body weight reduction for males appears permanent, i.e., there is a reduced set-point for body weight regulation (Powley and Keesey, 1970). The effect upon body weight for females may not be permanent unless the LH damage is sustained during infancy (Almli et al., 1979). The reduced body weight following LH damage is primarily related to reduction of body fat stores; however, body water and fat-free solids are also reduced in these brain-damaged animals. Animals with LH destruction also display greater body weight reduction than would be predicted by the volume and quality of food intake (Morgane, 1961), and they display disrupted feeding and drinking patterns (Morrison, 1968).

Lateral hypothalamus-damaged animals appear to be functionally desalivated, and they display an increased incidence of gastric pathology and altered gastric motility. This neural damage is also associated with an increased core body temperature, increased creatinine excretion (Morrison, 1968), increased urine output, and increased content and secretion of adrenal medulla catecholamines (Nathan and Reis, 1975).

A reduction in metabolic rate is also associated with LH destruction. These brain-damaged animals display a reduction in insulin (Steffens et al., 1972) and decreased thyroid weights and secretion (Davis, 1977). The LH-damaged animals may also display elevated blood glucose levels (Gray, 1971). In addition, animals with LH destruction display disturbances in motor, sensory, and arousal function (Morrison, 1968).

Animals sustaining LH destruction display a wide variety of changes in autonomic, metabolic, and endocrine function. The diversity of the functional changes presented here is intended to be representative rather than exhaustive, and to emphasize the complexity of functional involvement of this neural region. This presentation shows that some of the LH-damaged induced functional changes are indeed similar to some of the functional changes associated with aging and aging-related pathology.

4.2. Ventromedial Hypothalamic Damage

Destruction of the ventromedial region of the hypothalamus (VMH) also results in a persistent pattern of changes in metabolic, autonomic, and endocrine function in a variety of species (adults). Some of the more outstanding changes that occur in animals following damage to the VMH are increased feeding (hyperphagia), decreased activity, and obesity (Hetherington, 1941). The obesity in these brain-damaged animals is due to tremendous accumulation of fat via adipocyte hypertrophy. Along with the fat accumulation, carcass fat-free solids and protein tend to be decreased (Montemurro and Stevenson, 1957), and these animals display a reduction in body water.

Although hyperphagia is frequently associated with the obesity of VMH-damaged animals, the obesity is not totally dependent upon overeating. Animals with VMH damage may become relatively obese even when food intakes are controlled (e.g., pair-feeding, Brobeck et al., 1943), and some animals become obese without overeating in a free-feeding situation. These results and the finding that brain-damaged animals lose body weight more slowly than control animals under food restriction suggest that the obesity of VMH-damaged animals is due at least in part to metabolic changes. Further support for this position is obtained from research on young rats. Infant rats sustaining VMH damage become obese without hyperphagia (Fisher et al., 1978; Hill et al., 1981).

Related to this issue, animals with VMH damage display elevation in insulin levels (Hales and Kennedy, 1964) and increased glucose (Steffens et al., 1972). In addition, these animals display increased gluconeogenesis (Goldman and Bernardis, 1975), decreased glucose tolerance, increased plasma glucagon, and increased deamination of amino acids (Goldman and Bernardis, 1975). Animals with VMH damage also show increases in lipogenesis, blood lipids, plasma fatty acids, cholesterol (Brobeck et al., 1943), and blood triglycerides (Frohman et al., 1969).

Other changes in the VMH-damaged animal include an increase in size of the gasrointestinal tract; increased weight of the heart, liver, and kidneys (Brobeck et al., 1943); increased size of pancreatic islets (Kennedy and Parker, 1963); and increased thyroid gland weight with decreased thyroid gland secretion (Hinman and Griffith, 1973). These brain-damaged animals also display an increase in pericardial and perirenal fat deposits, as well as gonadal atrophy and dysfunction (Kennedy, 1969). Further, VMH-damaged animals have decreased levels of growth hormone, altered corticosterone output, increased secretion of gastric acid, and increased plasma urea and nitrogen excretion. Animals with VMH damage tend to be finicky feeders and they prefer high-fat content diets to standard diets (Brobeck *et al.*, 1943). These animals also display altered feeding patterns and increased reactivity to sensory stimulation.

From the above presentation of representative changes produced in animals by VMH damage, it is obvious that these animals display widespread changes in metabolic, autonomic, and endocrine function. Many of these brain-damage-induced functional changes resemble some of the metabolic, autonomic, and endocrine changes associated with the aging process, and especially, aging-related pathology.

5. AGING–HYPOTHALAMIC DAMAGE COMPARISONS

From the above presentation it is clear that damage to the lateral or ventromedial hypothalamus results in a wide variety of changes in metabolic, autonomic, and endocrine function. These results are of interest for two reasons. First, the effects of damage to these two hypothalamic regions are often (though not always) in opposite directions. And second, many (but not all) of the different types of dysfunction seen after hypothalamic damage are also frequently seen in aging and aging-related pathology.

A comparison of the effects of LH damage with VMH damage reveals many forms of dysfunction that are in opposite directions. For example, VMH damage results in *increased* body weight, *increased* carcass fat, *increased* insulin, and *increased* metabolic rate, while LH damage results in *decreased* body weight, *decreased* carcass fat, *decreased* insulin, and *decreased* metabolism. However, all forms of dysfunction are not in opposite directions following damage to these hypothalamic regions. For example, both LH and VMH destruction result in increased gastric acid secretion, decreased thyroid secretion, and attenuated locomotor activity levels. It is unfortunate that there is so little research specifically comparing the physiological effects of damage of these two hypothalamic regions.

With respect to aging and aging-related pathology, a comparison of the effects of hypothalamic damage and the aging process yields some very interesting similarities. Such a comparison reveals that many of the aging-related changes in behavior and physiology can be experimentally mimicked or approximated with experimental hypothalamic damage. For example, LH destruction results in a variety of functional alterations that are often seen during aging and aging-related pathology such as a decrease in body weight, decreased body water and fat-free solids, decreased metabolism, altered appetite, increased creatinine excretion (skeletal muscle wasting), increased urine output, decreased thyroid weight and secretion, and attenuated locomotor activity level. Many of the changes seen following LH damage appear to resemble the involution or atrophy pattern of aging (see Groen, 1959). However, many of the changes seen during aging and those seen following LH damage do not strictly correspond, e.g., decreased body fat stores with LH damage.

Damage to the VMH revealed even more similarities with aging and aging-related pathology than was found for LH damage. This, however, may be partly due to the greater volume of research on the physiological effects of VMH damage. As often associated with aging and aging-related pathology, VMH damage results in increased body fat stores, decreased body water and fat-free solids, altered appetite, increased insulin, decreased thyroid secretion, increased lipids in fat cells, disrupted glucose tolerance, increased lipogenesis, increased cholesterol, increased heart weight, gonadal atrophy and dysfunction, decreased growth hormone, and altered corticosterone output. Many of the effects of VMH damage tend to resemble the obesity pattern of aging (see Groen, 1959).

Groen (1959) has proposed that aging-related changes within the hypothalamus (cumulative lesions?) may become manifest in two different basic patterns of aging. One basic pattern of hypothalamic change is associated with hypometabolism and involution, and this pattern of aging may be associated with relative longevity. The other basic pattern of hypothalamic change is associated with hypermetabolism and obesity, and this pattern of aging is frequently associated with early onset of aging-related pathology (e.g., diabetes, hypertension) and a relatively decreased lifespan. This dichotomy is reminiscent of the suggestive relation between somatic morphology ("constitution") and the manifestation of disease states (Bauer, 1944). For example, the pyknic constitution is often associated with hypertension, elevated serum lipid levels, and coronary heart disease, while the leptosomic constitution is more often associated with gastritis and intestinal ulcers. The hypothalamic lesion research presented above also results in a relative dichotomy between LH and VMH animals based upon somatic morphology. Further, the metabolic, autonomic, and endocrine changes associated with experimental hypothalamic lesions are in the direction appropriate to the animal's new or altered "constitution."

At the risk of overgeneralization, it is tempting to speculate that there may be two "basic" or "extreme" patterns of aging and/or aging-related pathology: involution and obesity. The involution pattern of aging may be associated with relative longevity, while the obesity pattern may be associated with relatively early death due to an acceleration of aging-related pathology. Both of these patterns of somatic morphology can be produced via manipulations of metabolism, e.g., excessive feeding or feeding restriction (McCay et al., 1935). In addition, these basic somatic morphologies can also be produced with selective hypothalmic damage. These parallels may indicate that the ultimate effects of hypothalamic damage on the aging process and aging-related pathology may be mediated through the hypothalamic regulation of metabolism. Unfortunately, there is no research on life-span and aging-related pathology for animals with metabolism altered by LH and VMH hypothalamic damage. However, based upon our limited knowledge of the relation between feeding, metabolism, and longevity (or delayed pathology), one might predict that animals sustaining LH damage would display greater longevity and a different pattern of aging-related pathology than those animals sustaining VMH damage.

With regard to this speculation, we (Almli *et al.*, 1979; Hill *et al.*, 1981) have been inflicting damage on the LH and VMH of *newborn* rats. In spite of the fact that the rats were sacrificed relatively early (approximately 200–250 days of age), there were obvious differences between the brain-damaged groups. The VMH-damaged rats that survived to the age of sacrifice were "old-looking." They were very obese and relatively inactive and their fur was yellow and coarse. Many of these rats died before the age of planned sacrifice, and death was often associated with respiratory problems. Postmortem examination revealed tremendous fat deposits (including pericardial fat deposits) in these animals, and they frequently displayed tumors of the body and the brain. In contrast, newborn rats sustaining LH damage "aged" more gracefully. These rats were trim and active and their fur appeared more youthful. Very few of these rats died before the time of planned sacrifice, and very few incidences of brain or body tumors were found. While these observations are made after the fact, the results, as preliminary as they are, appear to be in the predicted direction.

The hypothalamic theories of aging-related changes in metabolic, autonomic, and endocrine function gain at least cautious support from research on the effects of experimental hypothalamic damage. This, of course, would be expected because the hypothalamic theories of aging have their foundation in the similarity between aging-related functional changes and the clinical manifestation of hypothalamic disease. It would be more surprising if hypothalamic lesions did not produce effects resembling hypothalamic disease. Nevertheless, it is becoming apparent that there may very well be an important relation among the hypothalamus, metabolism, and the aging process, especially with regard to aging-related pathology. The problem, of course, remains the determination of what is "cause" and what is "effect."

Groen (1959) has suggested that an important factor in the aging process is the accumulation of lesions (hypothalamic?) that commences at the arrest of cell division. The effects of these accumulating lesions may become clinically manifest as aging advances. It is possible that such "lesions" may be responsible for the "hypothalamic disregulation" proposed by Frolkis (1972, 1976) and the "altered hypothalamic thresholds" proposed by Dilman (1976, 1979). The effects of gross hypothalamic damage presented earlier would certainly be expected to produce hypothalamic disregulation and/or altered hypothalamic thresholds. However, it would not be expected that "typical" aging would be associated with such gross hypothalamic damage.

In light of recent research, however, it may not be farfetched to entertain the notion that hypothalamic "lesions" may underlie some of the metabolic, autonomic, and endocrine changes associated with aging and aging-related pathology. For example, Sabel and Stein (1981) have found significant neuronal shrinkage and neuronal loss within the lateral and ventromedial hypothalamic regions of rats during aging. These degenerative changes were found to be cumulative throughout aging. Thus, it is conceivable that degenerative changes of the hypothalamus may reflect the product of a genetic program–environment interaction, and such degenerative changes may represent the "accumulating lesions" proposed by Groen (1959). Similar to aging-related changes in the structure of other brain regions, aging-related structural changes within the hypothalamus would be expected to be associated with alterations in function. Degenerative changes of the hypothalamus would tend to produce functional changes of a metabolic, autonomic, and/or endocrine nature.

6. OVERVIEW AND SUMMARY

In this chapter a representative sample of the metabolic, autonomic, and endocrine changes associated with aging and aging-related pathology were presented. These aging-related functional changes influenced Groen, Dilman, and Frolkis to propose hypothalamic theories of aging based upon the well-documented hypothalamic involvement in the regulation of metabolic, autonomic, and endocrine function.

According to the hypothalamic theories of aging, some form of hypothalamic dysfunction or change is responsible for many of the manifestations of aging and aging-related pathology. To evaluate these theories, a selective review of the effects of experimental hypothalamic damage was presented. This review showed that the hypothalamus is affected in a variety of disease states, and that hypothalamic damage is associated with alteration in immune system function and changes in metabolic, autonomic, and endocrine function. Many of these brain-damage effects are similar to the alterations in function often associated with the aging process and aging-related pathology. In addition, it has been shown that the hypothalamus may in fact display aging-related degenerative changes, and it is conceivable that these hypothalamic degenerative changes could be responsible for some of the aging-related changes in metabolic, autonomic, and endocrine function, and thus the changes associated with aging-related pathology. What is missing from this hypothalamic–aging relation is an indication of what is "cause" and what is "effect."

More research on this topic is sorely needed, such as morphological, neurochemical, and electrophysiological analysis of the hypothalamus during development, maturity, and aging (especially cases of "premature" aging). In addition, it is important to determine the effects of cumulative hypothalamic lesions, and other manipulations of metabolism, on life-span and aging-related pathology. While the available data are suggestive of a role for the hypothalamus in the regulation of aging-associated changes in metabolic, autonomic, and endocrine function, the evidence for this role is primarily circumstantial at this time.

In spite of the difficulty in conducting aging research, the relation between the hypothalamus and the aging process merits our attention. There is a growing body of research that indicates that hypothalamic regulation of metabolic, autonomic, and endocrine function changes during early development from birth through maturity (Almli, 1978), and there is evidence that such regulation also changes during aging (Frolkis, 1976). Thus, if brain changes (e.g., growth) during early development may underlie changes in function, it is not a giant step to suggest that brain changes (e.g., cell atrophy, altered neurochemistry) may also be the basis of functional changes associated with aging. The hypothalamus may thus be an important neural region upon which to focus our research attention.

In summary, the growing evidence indicates that the hypothalamus may play a role in the regulation of some aspects of aging and aging-related pathology. However, while the nature of this role (cause or effect) is unknow at this time, the role played by the hypothalamus in the aging process will ultimately be a function of its interrelations with other neural areas and the remainder of the body. It is to be hoped that the hypothalamus will not become an "aging center" (à la "feeding center"), but will become better known as to its role in the regulation of aging and aging-related pathology.

REFERENCES

Almli, C. R., 1978, The ontogeny of feeding and drinking: Effects of early brain damage, *Neurosci. Biobehav. Rev.* **2**:281–300.

Almli, C. R., and Weiss, C. S., 1975, Behavioral and physiological responses to dipsogens: A comparative analysis, *Physiol. Behav.* **14**:633–641.

Almli, C. R., Hill, D. L., McMullen, N. T., and Fisher, R. S., 1979, Newborn rats: Lateral hypothalamic damage and consummatory-sensorimotor ontogeny, *Physiol. Behav.* **22**:767–773.

Ban, T., 1975, Fiber connections in the hypothalamus and some autonomic functions, *Pharmacol. Physiol. Behav.* **3**(Suppl. 1):3–13.

Bauer, J., 1944, *Constitution and Disease*, Heinemann, London.

Berlin, M., and Wallace, R. B., 1976, Aging and the central nervous system, *Exp. Aging Res.* **2**:125–164.

Bourne, G. H., 1973, Lipofuscin, *Prog. Brain Res.* **40**:187–201.

Brobeck, J. R., Tepperman, J., and Long, C. N. H., 1943, Experimental hypothalamic hyperphagia in the albino rat, *Yale J. Biol. Med.* **15**:831–853.

Brody, H., 1973, Aging of the vertebrate brain, in: *Development and Aging in the Nervous System* (M. Rockstein, ed.), Academic Press, New York, pp. 131–134.

Comfort, A., 1964, *Aging, the Biology of Senescence*, Holt, Rinehart and Winston, New York.

Davis, J. R., 1977, Decreased metabolic rates contingent upon lateral hypothalamic lesion-induced body weight losses in male rats, *J. Comp. Physiol. Psychol.* **91**:1019–1031.

Dilman, V. M., 1976, The hypothalamic control of aging and age associated pathology: The elevation mechanism of aging, in: *Hypothalamus, Pituitary and Aging* (A. V. Everitt and J. A. Burgess, eds.), Charles C. Thomas, Springfield, Illinois, pp. 634–667.

Dilman, V. M., 1979, Hypothalamic mechanisms of aging and of specific age pathology. V. A model for the mechanism of human specific age pathology and natural death, *Exp. Gerontol.* **14**:287–300.

Everitt, A. V., 1976a, The nature and measurement of aging, in: *Hypothalamus, Pituitary and Aging* (A. V. Everitt and J. A. Burgess, eds.), Charles C. Thomas, Springfield, Illinois, pp. 5–42.

Everitt, A. V., 1976b, The thyroid gland, metabolic rate and aging, in: *Hypothalamus, Pituitary and Aging* (A. V. Everitt and J. A. Burgess, eds.), Charles C. Thomas, Springfield, Illinois, pp. 511–528.

Everitt, A. V., 1976c, Conclusion: Aging and its hypothalamic–pituitary control, in: *Hypothalamus, Pituitary and Aging* (A. V. Everitt and J. A. Burgess, eds.), Charles C. Thomas, Springfield, Illinois, pp. 676–702.

Finkelstein, J. W., Roffwaig, H. P., Boyar, H. M., Kream, J., and Hellman, L., 1972, Age related changes in the twenty-four-hour spontaneous secretion of growth hormone, *J. Clin. Endocrinol.* **35**:665–670.

Fisher, R. S., and Almli, C. R., 1979, Postnatal ontogeny of hypothalamic extracellular unit activity in the rat, *Neurosci. Abstr.* **5**:159.

Fisher, R. S., Almli, C. R., and Parsons, S., 1978, Infant rats: Ventromedial hypothalamic damage and the development of obesity and neuroendocrine dysfunction, *Physiol. Behav.* **21**:369–382.

Frohman, L. A., Bernardis, L. L., Schnatz, J. D., and Burek, L., 1969, Plasma insulin and triglyceride levels after hypothalamic lesions in weanling rats, *Am. J. Physiol.* **216**:1496–1501.

Frolkis, V. V., 1976, The hypothalamic mechanisms of aging, in: *Hypothalamus, Pituitary and Aging* (A. V. Everitt and J. A. Burgess, eds.), Charles C. Thomas, Springfield, Illinois, pp. 614–633.

Froklis, V. V., Bezrukov, V. V., Duplenko, Yu. K., and Genis, E. D., 1972, The hypothalamus in aging, *Exp. Gerontol.* **7**:169–184.

Goldman, J. K., and Bernardis, L. L., 1975, Gluconeogenesis in weanling rats with hypothalamic obesity, *Horm. Metab. Res.* **7**:148–152.

Gray, R. H., 1971, The effects of bilateral destruction of the hypothalamic feeding center on fasting blood sugar, glucose absorption and glucose tolerance in the rat, *Aust. J. Exp. Biol. Med. Sci.* **49**:225–232.

Groen, J. J., 1959, General physiology of aging, *Geriatrics* **14**:318–331.

Gutstein, W. H., Harrison, J., Parl, F., Kiu, G., and Avitable, M., 1978, Neural factors contribute to atherogenesis, *Science* **199**:449–451.

Hales, C. N., and Kennedy, G. C., 1964, Plasma Glucose, non-esterified fatty acid and insulin concentrations in hypothalamic-hyperphagic rats, *Biochem. J.* **90**:620–624.

Hetherington, A. W., 1941, Th relation of various hypothalamic lesions to adiposity and other phenomena in the rat, *Am. J. Physiol.* **140**:89–92.

Hill, D. L., Almli, C. R., Fisher, R. S., and Williams, D., 1981, VMH damage in newborn rats: Growth, ingestion and neuroendocrine dysfunction, *Exp. Neurol.* **71**:191–202.

Hinman, D. J., and Griffith, D. R., 1973, Effects of ventromedial hypothalamic lesions on thyroid secretion rate in rats, *Horm. Metab. Res.* **5**:48–50.

Hirano, A., Arumugasamy, N., and Zimmerman, H. M., 1967, Amyotrophic lateral sclerosis: A comparison of Guam and classical cases, *Arch. Neurol.* **16:**357–363.

Kennedy, G. C., 1969, The relation between the central control of appetite, growth and sexual maturation, *Guys Hosp. Rep.* **118:**315–327.

Kennedy, G. C., and Parker, R. A., 1963, The islets of Langerhans in rats with hypothalamic obesity, *Lancet* **2:**981–982.

Keys, A., 1952, The age trend of serum cholesterol and of Sf 10-20 ("G") substance in adults, *J. Gerontol.* **7:**201–206.

Knigge, K. M., Hoffman, G. E., Joseph, S. A., Scott, D. E., Sladek, C. D., and Sladek, J. R., Jr., 1980, Recent advances in structure and function of the endocrine hypothalamus, in: *Handbook of the Hypothalamus,* Volume 2: *Physiology of the Hypothalamus* (P. J. Morgane and J. Panksepp, eds.), Marcel Dekker, New York, pp. 63–164.

Korneva, E. A., 1976, Neurohumoral regulation of immunological homeostasis, *Hum. Physiol.* **2:**374–384.

McCay, C. M., Crowell, M. F., and Maynard, L. A., 1935, The effect of retarded growth upon the length of the life span and ultimate body size, *J. Nutr.* **10:**63.

Miller, A. E., Shaai, C. J., and Riegle, G. D., 1976, Aging effects on hypothalamic dopamine and norepinephrine content in the male rat, *Exp. Aging Res.* **2:**475–480.

Montemurro, D. G., and Stevenson, J. A. F., 1957, Body composition in hypothalamic obesity derived from estimations of body specific gravity and extracellular fluid volume, *Metabolism* **6:**161–168.

Morgane, P. J., 1961, Medial forebrain bundle and "feeding centers" of the hypothalamus, *J. Comp. Neurol.* **117:**1–25.

Morgane, P. J., and Panksepp, J. (eds.), 1979, *Handbook of the Hypothalamus,* Volume 1: *Anatomy of the Hypothalamus,* Marcel Dekker, New York.

Morgane, P. J., and Panksepp, J. (eds.), 1980a, *Handbook of the Hypothalamus,* Volume 2: *Physiology of the Hypothalamus,* Marcel Dekker, New York.

Morgane, P. J., and Panksepp, J. (eds.), 1980b, *Handbook of the Hypothalamus,* Volume 3A: *Behavioral Studies in the Hypothalamus,* Marcel Dekker, New York.

Morgane, P. J., and Panksepp, J. (eds.), 1981, *Handbook of the Hypothalamus,* Volume 3B: *Behavioral Studies of the Hypothalamus,* Marcel Dekker, New York.

Morrison, S. D., 1968, The relationship of energy expenditure and spontaneous activity to the aphagia of rats with lesions in the lateral hypothalamus, *J. Physiol.* **197:**325–343.

Nathan, M. A., and Reis, D. J., 1975, Fulminating arterial hypertension with pulmonary edema from release of adrenomedullary catecholamines after lesions of the anterior hypothalamus in the rat, *Circ. Res.* **37:**226–235.

Oomura, Y., 1976, Significance of glucose, insulin, and free fatty acid on the hypothalamic feeding and satiety neurons, in: *Hunger: Basic Mechanisms and Clinical Implications* (D. Novin, W. Wyrwicka, and G. A. Bray, eds.), Raven Press, New York, pp. 145–157.

Ordy, J. M., and Kaack, G., 1975, Neurochemical changes in composition, metabolism and neurotransmitters in the human brain with age, in: *Neurobiology of Aging* (J. M. Ordy and K. R. Brizzee, eds.), Plenum Press, New York, pp. 253–285.

Papez, J. W., 1937, A proposed mechanism of emotion, *Arch. Neurol. Psychiatr.* **38:**725–743.

Paunovic, V. R., Petrovic, S., and Jankovic, B. D., 1976, Influence of early postnatal hypothalamic lesions on immune responses of adult rats, *Period. Biol.* **78:**50–57.

Pearl, R., 1905, Biometrical studies on man. I. Variation and correlation in brain weight, *Biometrika* **4:**13–104.

Pfaff, D. W., 1981, Theoretical issues regarding hypothalamic control of reproductive behavior, in: *Handbook of the Hypothalamus,* Volume 3B: *Behavioral Studies of the Hypothalamus* (P. J. Morgane and J. Panksepp, eds.), Marcel Dekker, New York, pp. 241–258.

Potvin, A. R., Syndulko, K., Tourtellotte, W. W., Lemmon, J. A., and Potvin, J. H., 1980, Human
    neurologic function and the aging process, *J. Am. Geriatr. Soc.* **28**:1–8.
Powley, T. L., and Keesey, R. E., 1970, Relationship of body weight to the lateral hypothalamic
    feeding syndrome, *J. Comp. Physiol. Psychol.* **70**:25–36.
Ranson, S. W., and Magoun, H. W., 1939, The hypothalamus, *Ergeb. Physiol.* **41**:56–163.
Sabel, B. A., and Stein, D. G., 1981, Extensive loss of subcortical neurons in the aging rat brain,
    Exp. Neurol. **73**:507–516.
Samorajski, T., 1977, Central transmitter substances and aging: A review, *J. Am. Geriatr. Soc.*
    25:337–348.
Scheibel, M. E., and Scheibel, A. B., 1977, Differential changes with aging in old and new cortices,
    in: *The Aging Brain and Senile Dementia* (K. Nandy and I. Sherwin, eds.), Plenum Press, New
    York, pp. 39–58.
Steffens, A. B., Mogenson, G. J., and Stevenson, J. A. F., 1972, Blood glucose, insulin and free
    fatty acids after stimulation and lesions of the hypothalamus, *Am. J. Physiol.* **222**:1446–1452.
Stein, M., Schiavi, R. C., and Camerino, M., 1976, Influence of brain and behavior on the immune
    system, *Science* **191**:435–440.
Young, C. M., Bloudin, J., Tensuan, R., and Fryer, J. H., 1963, Body composition studies of older
    women, thirty to seventy years of age, *Ann. N.Y. Acad. Sci.* **110**:589–607.

3

Morphological Measurements in the Aging Rat Cerebral Cortex

MARIAN C. DIAMOND and JAMES R. CONNOR

1. INTRODUCTION

This chapter deals with the structural changes seen in the brains of aging rats exposed to differential environments. We have previously reviewed the overall progress of experiments from our laboratory indicating brain changes due to modified external stimuli (Diamond and Connor, 1981, 1982). We showed changes in occipital cortical cell number and pyramidal cell dendrites from animals living in enriched, standard colony, and impoverished conditions. The present chapter includes some of the same data, but adds new information from the somatosensory cortex which has more recently been obtained. New dendritic spine data have also been presented to compare with our previous work.

One of the major problems in studying the aging process is determining the sequential order of the patterns of transition: If one area is aging more rapidly than another, how does the first change affect other areas? A specific question to ask concerning the nervous system is what are the factors influencing the aging process of the nondividing nerve cell. Are they primarily the condition of the support systems to the nervous system such as the cardiovascular system, or the status of the alveolar epithelium of the lung, or the degradation of any other organ of the body which directly or indirectly affects neural function? If all the support systems would be maintained in optimal condition, would nerve cells have the potential to function endlessly?

MARIAN C. DIAMOND and JAMES R. CONNOR • Department of Physiology–Anatomy, University of California, Berkeley, California 94720. This chapter is dedicated to the memory of Dorothy E. McClanahan.

If we narrow down our thoughts still more specifically to the brain and then in turn to the cerebral cortex, we have learned that certain regions of the cortex age before others. For example, it is reported that as age advances the tertiary fields show the first signs of alteration, i.e., a decrease in overall myelin density, whereas the core fields remain well preserved for a longer time (Yakolev and Lecours, 1967). If such differential aging patterns are observed in a horizontal direction, are similar observations to be found in a vertical direction through the layers of the cortex? If such events are noted, then another question which can be asked is whether the layers of the cortex can be differentially affected by the type of external environment in which the aging individual lives.

We have learned that purely qualitative morphology of old animals or old people in general no longer seems adequate in quantifying age tissues. Precise chronological age is slowly losing its meaning as a single determinate of the status of aging. Studies using a systematic, blind (in regard to the age of the animal or human) approach have found that the aging brain is not characterized solely as a phenomenon of deterioration (Hinds and McNelly, 1977; Buell and Coleman, 1979; Connor et al., 1980, 1982a). In our experiments with rats, we have made an initial attempt to control for some of the numerous variables which are constantly influencing neural function.

Our report will consider the changes in neuronal and glial numbers in the rat occipital cortex during aging and also the changes in pyramidal cell dendritic patterns in specific cortical layers in animals living in standard colony conditions. In addition, the aging pattern will be examined in cortical neurons from middle-aged and old animals in enriched environments and in old animals in impoverished environments. Some of the results have been presented previously and some are new. Both are reported here in an attempt to obtain a better understanding of the whole picture.

2. NEURONAL AND GLIAL CELL NUMBER CHANGES DURING AGING

Initial steps have been taken to understand cortical cell loss with aging. The study of Brizzee et al. (1968) showed that there was no significant loss of neurons in the rat somatesthetic cortex after 100 days of age to about 700 days of age. We counted cells in the occipital cortex (Diamond et al., 1977) in 26-, 108-, and 650-day-old rats (Long-Evans) which had been housed three to a cage (30×20×28 cm) well into old age. Then the very large, old animals were changed to two rats per cage. From 6-μm, celloidin-embedded tissue stained with luxol fast blue, cell counts were made on enlarged photographs with the final enlargement from the slide to the print being about 800×. The greatest decrease in density in neurons and in two types of glia, astrocytes and oligodendrocytes, occurred before 108 days of age.

There was no significant loss of neurons or glia after 108 days of age. Our most recent unpublished data show that even at 904 days of age there is no significant further loss of neurons and glia.

It is not clear from the decrease in number of neurons per unit area whether they are being redistributed or are actually dying. The area of the perikaryon was also found to decrease significantly between 41 and 108 days (10%, $P < 0.01$), but not between 108 and 650 days of age. The nucleus showed a similar decrease during this time, but the difference was not significant. In other words, the greatest changes in cell number and dimensions were taking place before 100 days of age when the animals were housed under standard colony conditions. As Minot (1908) says, "The period of old age, so far from being the chief period of decline, is in reality essentially the period in which the actual decline going on in each of us will be the least."

The cortex was divided equally into upper and lower halves from the beginning of layer II to the lower border of layer VI. (We have yet to plot the cell densities through the specific layers but will do so in the future.) From our present data we have learned that from 41 to 108 days of age, the neurons in the lower layers decrease more markedly, by 24% ($P < 0.001$), per unit area than in the upper layers, 9% ($P < 0.01$). Why the neurons projecting from the cortex are being reduced per unit area more readily than those in the integrative, receptive layers is not clear. It is possible that their neuropil is growing more rapidly, thus separating perikarya per unit area. For the comparison from 108 to 650 days of age, no significant differences in rates of neuron density decrease were found between upper and lower layers.

We found from our glial counts, including both astrocytes and oligo-dendrocytes, that the pattern of aging in our standard colony rats followed identically that of the neurons, both before 108 days of age and after this time, at least until 650 days of age, an old age with this group of our Long-Evans strain.

We were interested to learn if we could obtain differences in neuron number between rats living in environmentally enriched or impoverished environments (Diamond et al., 1966; Diamond, 1976). In counting neurons and glia in the same region, the occipital cortex, that we had studied for the standard colony rats, we found a decrease in number of neurons per unit area in the enriched animals compared with the impoverished (Diamond et al., 1966). However, this decrease did not reach a statistically significant level. At the same time the cortex was thicker in the enriched rats compared with the impoverished rats, indicating that the neuropil had increased in the enriched animals without significantly changing the neuron number. It is not certain whether the former neuron decrease per area seen in the standard colony animals was prevented by having the animals live in the enriched environment. Both glial cells and dendrites are greater in number in the enriched animals. We found that the glial cells, primarily the oligodendrocytes, were greater in number by 14% ($P < 0.01$) in the enriched compared with the impoverished animals. The increased number of glial cells could mean several things. For example, they could signify an increased neuropil which has been demonstrated first in our laboratory by Holloway (1966) or an increase in perikarya dimension which we have shown (Diamond, 1967). In order to understand the true picture of the effects of enriched environments on neuron number during the formative days of the postweaned cortex, we would have to start with genetically controlled triplets and count cells in one group at 26 days of age and then in the enriched group and standard colony groups after 108 days of age. In this experiment we would have adequate controls to understand the effects of environment on cell number.

2.1. Dendritic Branching

We were then interested in learning whether the dendritic loss reported by others (Scheibel *et al.*, 1975; Feldman, 1977, to name a few) was inevitable with the passage of time as the only significant variable. The object of our investigation of aging dendritic patterns was to learn first about their structure in the standard colony aging rats and then to learn how exposure to differential environments could alter this pattern (Connor *et al.*, 1980a,b, 1981a,b, 1982c).

The experimental design was divided into two parts. The first part consisted of taking 36 Long-Evans rats at 414 days of age and dividing them into three groups: 12 baseline (i.e., sacrificed at 414 days of age), 12 standard colony (SC), and 12 enriched (EC). Both of the latter groups were sacrificed at 444 days of age. In the second part ten animals remained in SC until 600 days of age. During this time three animals died, so only seven were available for the remainder of the experiment. Since 12 animals usually constitute the enriched group, it was necessary to redesign our experimental paradigm and place four 600-day-old animals with eight 60-day-old animals in the enriched condition. The remaining three 600-day-old animals lived in the standard condition in a small cage and three groups of three 60-day-old rats were also placed in standard colony cages.

After 30 days in their respective environments, all animals were killed. The brains were removed from sodium-nembutal-anesthetized rats and placed in a Golgi-Cox solution modified by Van der Loos (1959). One-hundred-and-fifty-micrometer-thick sections were cut from celloidin-embedded tissue. Basal dendrites from seven pyramidal cells per section were taken from layers II and III in areas 18, 17, and 18a for the occipital cortex. A total of 756 neurons was observed from the 414- to 444-day group and 504 neurons were drawn from the 60- to 90- and 600- to 630-day groups. There were no significant differences between the SC and EC dendritic measures so the data from these two groups were combined. In the somatosensory cortex (areas 4, 3, 1, 2), 252 neurons were traced for 414-day-old group, 504 for the 444-day-old group, and 147 at 630 days of age.

In the occipital cortex, the SC animals from the 414- to 444-day-old group had more first-order branches than in either the 90- or 630-day-old animals, whereas, the frequency of the third- and fourth-order dendrites was markedly higher for the 90- and 630-day-old animals than the 414- and 444-day-old animals. In the somatosensory cortex, the 630-day-old rats had more third- and sixth-order segments than the 414- and 444 day-old rats.

2.2. Dendritic Density

The density of the basal dendritic tree was analyzed on the neurons drawn at each of the age groups. This is a measure of both the number and length of the segments. To do this, a series of concentric circles are placed over the neuron tracing and each intersection of the branch and circle is counted. The data are discussed in terms of four ranges: 0–50, 50–100, 100–150, and 150–200 μm from the soma. In the occipital cortex, the range nearest the soma (0–50 μm) was nearly the same for all groups. In the subsequent ranges moving more distally from the soma, the 90- and 630-day-old groups remained similar while there was a statistically significant drop in the density of the basal dendritic tree of the middle-age group ($P < 0.001$) (Connor et al., 1981a,b).

In the somatosensory cortex, in addition to measuring the density of the basal dendritic tree, we determined the location of the density changes by utilizing an orientation factor. The concentric circle series was divided into an upper and lower sector by lines which created an angle of 240° in the upper sector and 120° in the lower sector (Connor et al., 1982a). Unlike in the occipital cortex, we found an increase in density in the 0- to 50-μm range in the somatosensory cortex. However, the orientation factor enabled us to pinpoint the density increase in the upper sector of this range. The lower sectors were not statistically different (Fig. 1). There were no statistically significant differences in the 50- to 100-μm range (Fig. 2). In the two most distal ranges, the 630-day-old group had a greater dendritic density in each of the sectors, as well as total density (Figs. 3 and 4). This finding of an increased density in the upper sector of the most proximal range has lead us to speculate that with age there is a loss of input into layer IV (either quantitative or qualitative), and the dendritic tree is shifting to search for an afferent supply. We have additional evidence for this in a recent study where we found animals that had been isolated during the last six months of their life had a less dense dendritic tree in the lower sector in layer III than animals which had been socially housed during the same period. These animals in this latter study were the same age as the ones in our previous investigations (Connor, 1981).

We have interpreted our findings of increased dendritic volume from middle to old age in rats as a compensatory increase or hypertrophy of existing neurons in response to a declining dendritic population in their vicinity. This notion was first proposed by Hinds and McNelly (1977) in the olfactory bulb. To prove this,

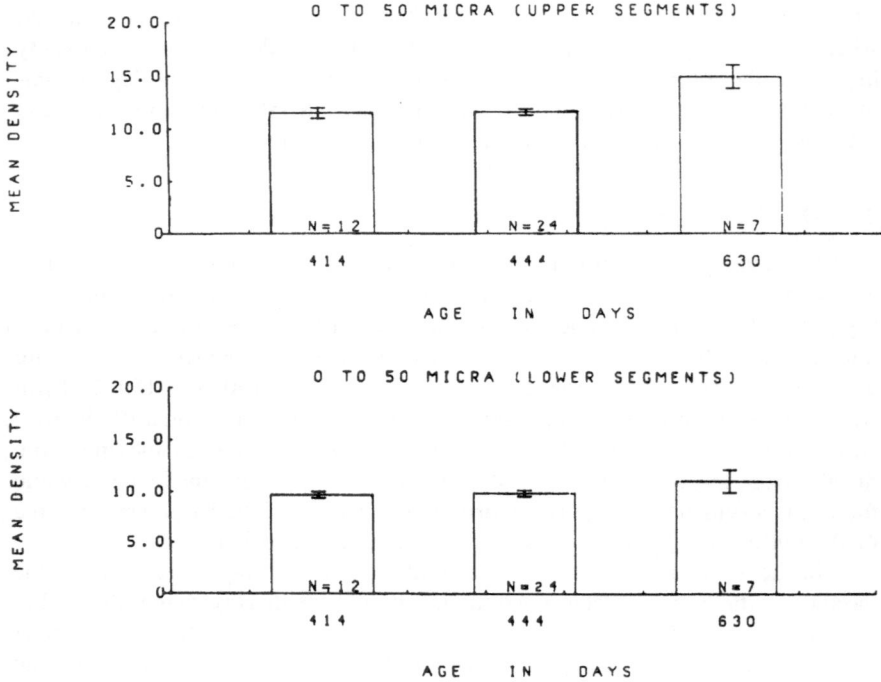

Figure 1.   The data for the range most proximal to the soma (0–50 μm) presented in upper and lower segments as the mean number of intersections ±SE.

we must show loss of dendritic material in the cortex. Diamond *et al.* (1977) reported a 4% loss of neurons in the occipital cortex with age. Although this percentage was statistically insignificant, its potential biological significance cannot be overlooked. That is, 4% of the neurons in 1 mm³ is 25 neurons. Conservatively estimating that each neuron had 15 segments, this figure would yield a loss of 375 dendritic segments from 1 mm³ of cortical tissue. Such a loss could not go unnoticed by the system.

In addition, in both the occipital and somatosensory cortices, we found a loss of primary (or first-order) segments with age. These are the segments which arise from the soma and give rise to the rest of the dendritic tree. Thus, even if a neuron would not compensate for loss of dendritic material in a neighboring cell it could compensate for its own dendritic loss within its own dendritic domain.

If our interpretation is accurate, it does not unequivocally follow that behavioral changes such as memory loss with age should not occur. One must realize that the dendritic tree is the primary source of integration of information in the central nervous system. Thus, a neuron which has undergone compensatory

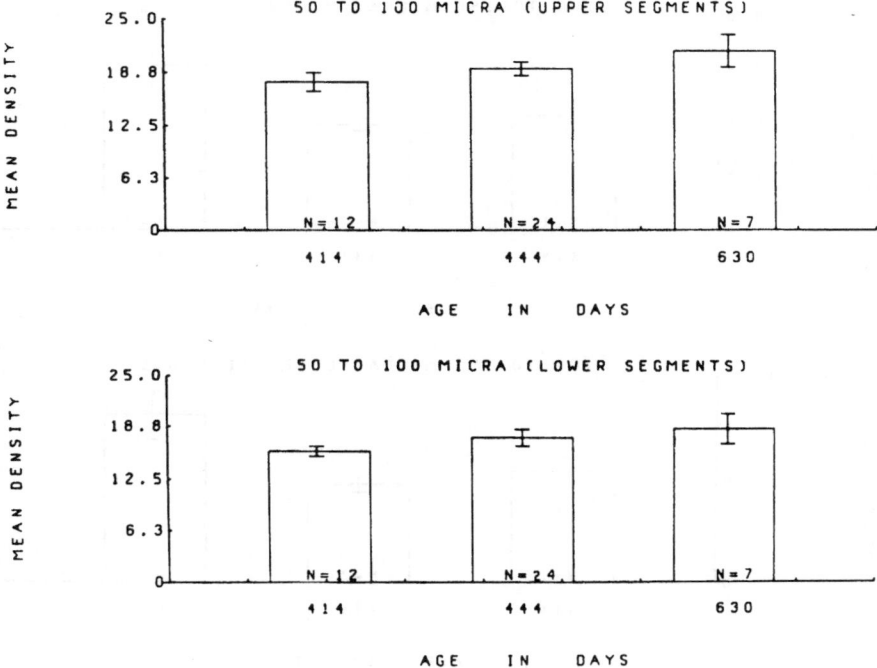

Figure 2. The second range was 50–100 μm from the soma. Data are presented as mean number of intersections ±SE.

hypertrophy of its dendritic tree has invited upon itself an increase in the electrophysiologic traffic it must assimilate.

On the other hand, in a study of parahippocampal gyrus of adult aged and aged demented humans, Buell and Coleman (1979) found that adult and aged demented individuals had the same amount of dendritic branching, whereas, nondemented elderly people had more branches. We interpret this evidence as support of the notion that compensatory hypertrophy of the dendritic tree can be functionally meaningful.

2.3. Dendritic Length

The animals that had spent their last 30 days in an enriched environment had longer terminal segments in the most distal portions of their dendritic trees in both the occipital and somatosensory cortices when compared to their SC littermates of the same age (Connor *et al.*, 1981b, 1982c). Specifically, the sixth-order terminal segments from EC animals were 86% longer than the sixth-order

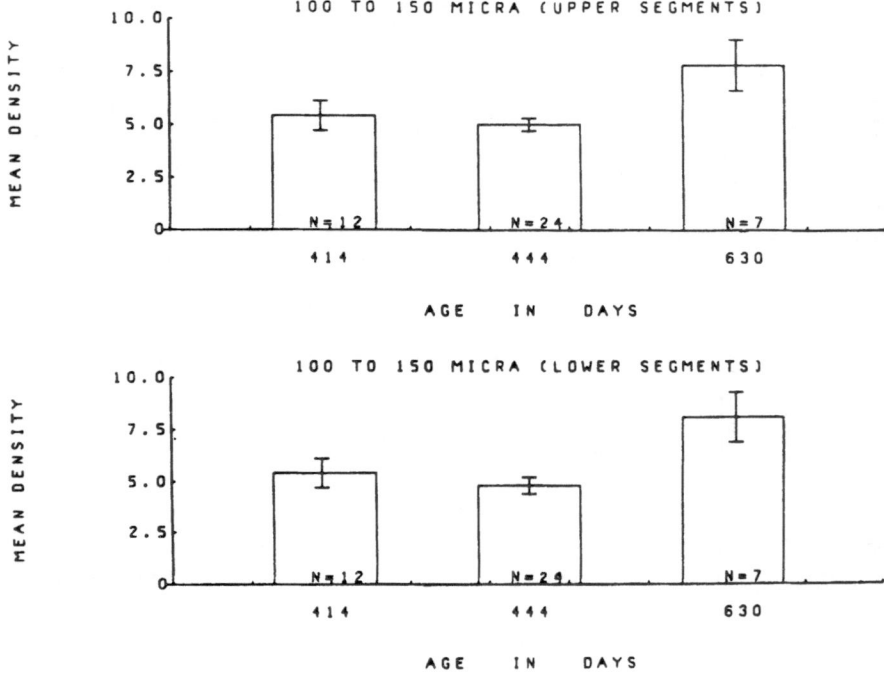

Figure 3. The third range from the soma (100–150 μm). Data presented as mean ±SE.

segments from SC rats in the occipital cortex. In the somatosensory cortex, the eighth-order segments were the most distal from the soma, and these segments were 80% longer on neurons from EC rats compared with their SC littermates. This is the first example of an environmental effect in aging animals. The response to the enriched environment may be interpreted either as a growth (elongation) of the distal segments or a retardation of retraction. Comparing our data with the available information causes us to favor the retardation of retraction as the more likely explanation at the present time.

2.4. Dendritic Spines

For determining the density of spines per unit length, the basal dendrites of the pyramidal cells of layers II and III from the occipital cortex were selected to correspond with our previous study (Connor *et al.*, 1980a). A 34-μm segment was counted beginning at the first bifurcation site. Two segments per neuron, one from each side of the soma, were chosen. Two types of spines were counted.

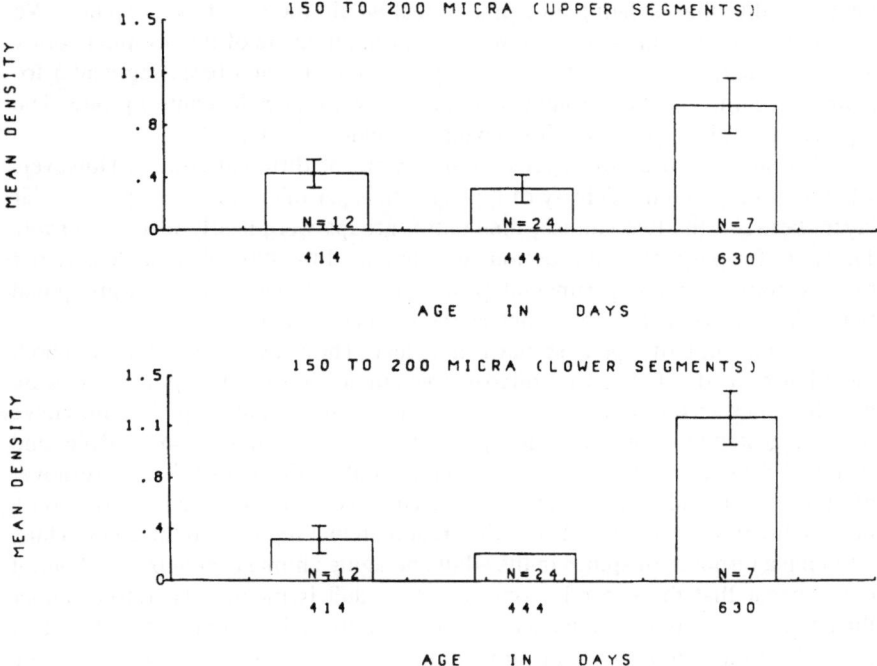

Figure 4. The most distal range from the soma (150–200 μm). Data are presented as the mean number of intersections ±SE.

Those with a stalk and a terminal bulb (lollipop configuration, type L) and those with no terminal expansion but a broad stalk (nubbin configuration, type N).

The type L spines and the total spines followed the same pattern as the dendritic branching with the higher spine density occurring at 90 and 630 days and with the lowest density in the middle-age groups, 414–444 days of age. The fluctuations in spine density were statistically significant. There were no significant differences in the results from animals living in the EC and SC within an age group. Because dendritic spines are the preferred site of synaptic contact, the data presented here suggest that the changes in dendritic branching we measured are incorporated into the neuropil as functional moieties.

The type N spines showed a different pattern of change than did the type L with age. The density of type N spines increased from 90 to 444 days ($P < 0.025$) and then remained relatively stable. There was a 33% decrease in the density of type N spines when the 630-day-old EC animals were compared with the 630-day-old SC animals.

Because of the decrease of type N spines in EC, we decided to investigate the possibility that isolating an aging animal would increase type N spines. We found that type N spines were more frequent in all layers of the occipital cortex and on all parts of the dendritic tree (apical, oblique, and basal segments) for animals from isolated environments compared with their SC control group. The type L spine displayed very little change (Connor, 1981).

The animals used in the two studies were of different strains. However, when we compared the density of type N spines per micron in the supragranular layers between the two studies, the control groups were nearly identical (Table 1). Thus, for purpose of discussion, we compared the type N spine density per micron from the two experimental groups and found that the isolated group had twice the density of type N spines as the enriched group.

The meaning of type N spines is not clear. They have been reported in both the adult rat and cat. Electron microscopic studies in the adult animals indicate that the synapses on these spines appear structurally normal. Type N spines have been suggested to be degenerating spines, but also developing spines (Miller and Peters, 1981). The developing spine notion would indicate that there is turnover of spines in the adult and aging animal. However, the synapse occurring on a developing type N spine is clearly different than that reported in the adult. Thus, although a turnover of spines in the adult and aging animal cannot be ruled out, it does appear that the synaptic contact in the adult is mature. Therefore, rather than consider the type N spine as developing in the aging animal, we interpret it as a reflection of a decrease or diminution of the afferent supply to the synaptic site.

Cortical thickness data were also collected on thionine-counterstained, Golgi-Cox sections adjacent to those used for dendritic measures. For all age groups, the EC cortical thickness was greater than that of the SC, but the

Table 1. Comparison of Mean Spine Density per Micron on Basal Dendritic Segments of Superficial Pyramidal Cells for 20-Month-Old Rats of Two Different Strains[a]

Condition[b]	Spine type[c]	Connor et al. (1980) (Long-Evans, Berkeley)	Connor and Diamond (1982) [GIBCO (Sprague-Dawley)]
IC	L	—	0.46
SC	L	0.45	0.45
EC	L	0.48	—
IC	N	—	0.12
SC	N	0.09	0.08
EC	N	0.06	—

[a]All data are from the visual cortex.
[b]IC, isolated condition; SC, standard colony (control); EC, enriched condition.
[c]See text for description.

differences were not statistically significant. In previous publications we have shown that the cortical thickness of rats living in enriched environments from 60 to 90 days of age showed statistically significantly greater thickness changes in the enriched animals compared with the standard colony (Diamond *et al.*, 1976). However, when the old rats were housed with the young, the young rats' brains did not develop the usual cortical thickness increases. We proposed that the old rats inhibited the young from interacting with the "toys." (Some people have interpreted these findings to suggest that one reason why professors stay young in mind as they age is because they constantly interact with their youthful students year after year.)

The evidence so far shows that by exposing old rats to enriched environments, the length of dendrites in the cerebral cortex is longer than in animals in standard colony conditions. Since the point we are making is that the environmental conditions to which the animal is exposed during aging are important, then it is essential to determine if the dendritic pattern can be altered by placing animals in cages by themselves in isolated conditions during aging. This experiment was the basis of Connor's Ph.D. thesis (1981). The lack of stimuli to young animals has been thoroughly demonstrated to decrease the dimensions of the cerebral cortex and its constituents (Diamond *et al.*, 1976). The question we are asking here is whether the old cortex is also responsive to stimulus deprivation or is morphologically stable during aging.

3. AGING AND IMPOVERISHMENT

A new group of experiments was designed using male, retired breeder, albino rats from GIBCO Animal Resources laboratory (formerly ARS/Sprague-Dawley) in Madison, Wisconsin. At 12 months of age the rats were separated three to a cage ($30 \times 20 \times 28$ cm). Then at 14 months of age they were randomly segregated so that some were isolated, one to a cage (isolated condition, IC), and the others remained three to a cage, the usual SC. All animals had water and food *ad libitum* and were on a 12-hr-light–12-hr-dark cycle. At 20 months of age, six rats from SC and six from IC were anesthetized and the brains removed. The Golgi-Cox-impregnated tissues were embedded in celloidin and sectioned at 150 μm. The primary visual area was used for further dendritic analysis.

3.1. Dendrites

The dendrities on pyramidal neurons from layers II and III were measured on Golgi-Cox-impregnated material. In addition, layer V was divided into upper and lower halves, Va and Vb, so that cells could be specially measured in these two divisions of layer V. The division of layer V was important because pyramidal neurons of layer Va are association neurons, and pyramidal neurons in layer

Vb are projection neurons. Only those neurons whose apical shaft extended beyond layer IV were drawn. Oblique branching, oblique length, and basal dendritic density were measured.

3.2. Oblique Branching

There were significantly more third-order terminal segments ($P < 0.05$) and total third-order segments on neurons ($P < 0.05$) from SC rats compared to those from IC rats in layer II. In layer III, the third-order intermediate segments and fourth-order terminal segments possessed a statistically significant difference, with neurons from SC rats having more than the IC rats ($P = 0.05$ in both cases).

The results from layer Va and Vb offered different findings in oblique branching. No statistically significant differences were observed for neurons in layer Va, although neurons from SC rats had a higher number of branches at each order compared with the IC rats. In layer Vb, however, rats' neurons had a greater frequency of second-order terminal segments ($P < 0.05$) than the neurons from SC rats. This was the only layer in which neurons from IC rats had more total oblique branches than neurons from SC rats.

3.3. Oblique Length

Both the first-order intermediate and first-order terminal segments were longer on neurons in layer II from SC rats compared with IC rats, with only the latter reaching a statistically significant level ($P < 0.008$).

In layer III, neurons from SC animals had significantly longer first-order intermediate segments in comparison with the IC. In layer Va there were no significant differences observed among the lengths of dendrites. However, in layer Vb, the neurons from the IC animals had significantly longer first-order intermediate segments ($P = 0.009$) than neurons from SC rats.

The oblique segments from neurons in layer III were more numerous than those of layer II, whereas, neurons from layer II had more basal segments than neurons from layer III. The influence of the housing conditions on the basal dendritic tree was less impressive. However, statistically significant ($P < 0.01$) shifts in the orientation of basal dendritic trees in layers II and III suggest that there were differences in the amount of afferent supply to the cortex between the two groups of experimental animals. The direction of shift of the dendritic tree is similar to that seen by Valverde (1970) in blinded and dark-reared rats. Valverde interpreted his finding as a searching out of an afferent supply to replace that which had been lost.

3.4. Spines

The salient finding in the spine density analysis was the consistent increase in density of spines which lacked a terminal expansion, the type N or nubbin spines, on segments of neurons from isolated rats compared with social rats. The greater density of type N spines was found irrespective of dendritic segment counted or cortical lamina. The functional significance of type N spines is not known, but they may represent an altered morphological state due to the functional input to the nerve cell.

Dendritic spines with a terminal expansion and resembling the lollipop (type L) spine were always present in a much greater frequency than type N spines. The type L spine density was not statistically different on neurons from rats in the two housing conditions except in layer Vb. In this layer the spines were more numerous on neurons from isolated rats than on neurons from social animals.

4. SUMMARY

The results we have presented indicate that during aging the changes occurring in the different layers are similar in some instances and not in others. The environment does affect the different layers in separate ways. By using only our standard colony environment we learned that there were greater decreases in the cell number per unit area in the lower half of the cortex compared with the upper half during the first 100 days of life. Thereafter, there were no such differences at least until 650 days of age.

The data on dendritic branching from both the occipital and somatosensory cortices reveal a pattern indicating that dendritic growth can occur in aging animals. The data on segment length show that the aging animal can respond to the external environment.

The dendritic spine analysis reveals a selective increase of type N spines with age and also with isolation. Thus, this spine type increases in density under conditions which are considered to have diminished afferent input.

These results continue to add information to the general pool indicating the importance of the role of the external environment in studying the aging nervous system and also the importance of considering more than one region of the brain at a time when mapping aging patterns.

REFERENCES

Brizzee, K. R., Sherwood, N., and Timiras, P. S., 1968, A comparison of various depth levels in cerebral cortex of young adults and aged Long-Evans rats, *J. Gerontol.* **23:**289–297.

Buell, S., and Coleman, P., 1979, Dendritic growth in the aged human brain and failure of growth in senile dementia, *Science* **206:**854–856.

Connor, J. R., Jr., 1981, Environmental influences on aging brain morphology, Ph.D. thesis, Department of Physiology–Anatomy, University of California, Berkeley, California.

Connor, J. R., and Diamond, M. C., 1982, A comparison of dendritic spine number and type on pyramidal neurons of the visual cortex of old adult rats from social or isolated environments, *J. Comp. Neurol.* **210:**99–106.

Connor, J. R., Diamond, M. C., and Johnson, R. E., 1980a, Occipital cortical morphology of the rat: Alteration with age and environment, *Exp. Neurol.* **68:**158–170.

Connor, J. R., Diamond, M. C., and Johnson, R. E., 1980b, Aging and environmental influence on two types of dendritic spines in the rat occipital cortex, *Exp. Neurol.* **70:**371–379.

Connor, J. R., Diamond, M. C., Connor, J. A., and Johnson, R. E., 1981a, A Golgi study of dendritic morphology in the occipital cortex of socially reared aged rats, *Exp. Neurol.* **73:**525–533.

Connor, J. R., Melone, J. H., Yuen, A. R., and Diamond, M. C., 1981b, Dendritic length in aged rats' occipital cortex: An environmentally induced response, *Exp. Neurol.* **73:**827–830.

Connor, J. R., Beban, S. E., Hopper, P. A., Hansen, B., and Diamond, M. C., 1982a, A Golgi study of the superficial pyramidal cells in the somatosensory cortex of socially reared old adult rats, *Exp. Neurol.* **76:**35–45.

Connor, J. R., Beban, S. E., Melone, J. H., Yuen, A., and Diamond, M. C., 1982b, A quantitative Golgi study in the occipital cortex of pyramidal dendritic topology of old adult rats from social or isolated environments, *Brain Res.* **251:**39–44.

Connor, J. R., Wang, E. C., and Diamond, M. C., 1982c, Increased length of terminal dendritic segments in old rats' somatosensory cortex: An environmentally induced response, *Exp. Neurol.* **78:**466–470.

Diamond, M. C., 1976, Anatomical brain changes induced by environment, in: *Knowing, Thinking, and Believing* (J. McGaugh and L. Petrinovich, eds.), Plenum Press, New York, pp. 215–241.

Diamond, M. C., and Connor, J. R., Jr., 1981, A search for the potential of the aging brain, in: *Brain Neurotransmitters and Receptors in Aging and Age-Related Disorders* (S. Enna, T. Samorajski, and B. Beer, eds.), Raven Press, New York, pp. 43–58.

Diamond, M. C., and Connor, J. R., 1982, Plasticity of the aging cerebral cortex, in: *Physiological and Pathophysiological Aspects of the Aging Brain* (S. Hoyer, ed.), Springer-Verlag, New York, pp. 36–44.

Diamond, M. C., Law, F., Rhodes, H., Lindner, B., Rosenzweig, M., Krech, D., and Bennett, E. L., 1966, Increases in cortical depth and glia numbers in rats subjected to enriched environments, *J. Comp. Neurol.* **128:**117–126.

Diamond, M. C., Johnson, R. E., and Gold, M. W., 1977, Changes in neuron and glia number in the young, adult, and aging rat occipital cortex, *Behav. Biol.* **20:**409–418.

Feldman, M. L., 1976, Aging changes in the morphology of cortical dendrites, in: *Neurobiology of Aging* (R. D. Terry and S. Gershon, eds.), Raven Press, New York, pp. 221–227.

Hinds, J. W., and McNelly, N. A., 1977, Aging of the rat olfactory bulb: Growth and atrophy of constituent layers and changes in size and number of mitral cells, *J. Comp. Neurol.* **171:**345–369.

Holloway, R. L., 1966, Increased dendritic branching in layer II stellate neurons of occipital cortex as compared to deprived littermates, *Brain Res.* **2:**393–396.

Miller, M., and Peters, A., 1981, Maturation of rat visual cortex. II. Electron microscopic study of pyramidal neurons, *J. Comp. Neurol.* **203:**555–573.

Minot, C. S., 1908, *The Problem of Age, Growth, and Death*, G. P. Putnam's Sons, New York.

Scheibel, M. E., Lindsay, R. D., Tomiyasu, U., and Scheibel, A. B., 1975, Progressive dendritic changes in aging human cortex, *Exp. Neurol.* **47:**392–403.

Valverde, F., 1970, The Golgi method: A tool for comparative structural analyses, in: *Contemporary Research Methods in Neuroanatomy* (W. J. H. Nauta and S. O. E. Ebbesson, eds.), Springer-Verlag, New York, pp. 12–28.

Van der Loos, H., 1959, *Dendro-dendritische Verbindingen in de Schors der Grote Hersenen*, H. Stam, Haarlem.

Yakovlev, P. I., and Lecours, A. R., 1967, The myelogentic cycles of regional maturation of the brain, in: *Development of the Brain in Early Life* (A. Minkowski, ed.), Blackwell Scientific Publications, Oxford, Edinburgh, pp. 3–70.

4

Morphological Aspects of Brain Damage in Aging

STEPHEN W. SCHEFF, KEVIN ANDERSON, and
STEVEN T. DeKOSKY

1. INTRODUCTION

The field of neurobiology has recently witnessed a surge of research aimed at studying and evaluating neural growth following various types of damage in the mammalian central nervous system (CNS). This research has taken on a very intriguing atmosphere since it has been suggested that this growth might serve as a basis for the recovery or restitution of function following damage. Conversely, one possibility must be recognized that aberrant neural circuitry formed as a result of remodeling may be functionally deleterious. Anomalous growth may thus be in part responsible for altered behavioral responses seen within aging. Interest in this area of research is further enhanced when one considers that this reactive growth process takes place not only in the developing nervous system but in the mature adult CNS.

    Over the past several years, a number of studies have shown that selective lesions which result in the loss of part of the input to a neuron cause the remaining undamaged inputs to sprout and form new synaptic connections in

STEPHEN W. SCHEFF • Departments of Anatomy and Neurology, University of Kentucky College of Medicine, Lexington, Kentucky 40536.   KEVIN ANDERSON • Department of Anatomy, University of Kentucky College of Medicine, Lexington, Kentucky 40536.   STEVEN T. DeKOSKY • Department of Neurology, Lexington Veterans Administration and University of Kentucky Medical Centers, and Sanders–Brown Research Center on Aging, Lexington, Kentucky 40536.

place of those lost (Cotman et al., 1981). In this way the circuitry of the brain is remodeled when it is damaged. This process of remodeling has been called axon sprouting, or more correctly, reactive synaptogenesis (Cotman and Lynch, 1976). Studies using young adult animals have shown that the growth process follows very strict temporal parameters and proceeds with a high degree of specificity. Our own work has shown that, in comparison to young adult animals, aged rats have a diminished capacity to support axon sprouting and hence fail to adequately reinnervate a massively denervated zone. The reasons for this decline or diminished capacity to reinnervate remain unclear. They may be associated with changes in hormonal levels, specific neurochemical pathways, and/or changes in neuronal membrane properties which are altered during the aging process. The fact remains that this diminished growth capacity is a feature of senescence in the CNS.

Age-related differences in reactive synaptogenesis may have functional significance (Steward, 1982). In some instances, lesion-induced growth appears to underlie recovery or retention of normal function after damage to the CNS (Dieringer and Precht, 1977; Goldberger and Murray, 1974; Loesche and Steward, 1977; Murray and Goldberger, 1974; Scheff and Cotman, 1977; Steward et al., 1977). In the aged brain we would expect this process of reactive synaptogenesis to operate not only in cases of severe damage, such as that induced in lesion studies, but also in the replacement of connections lost as a result of neuronal death accompanying the natural aging process. If the new connections can functionally replace the old, then reactive synpatogenesis may be regarded as a compensatory mechanism that counteracts the effects of aging. A reduction in growth capacity with age may therefore be detrimental and be partially responsible for age-related changes observed in behavior. Reliable age-related cognitive deficits have been documented in carefully controlled studies of many mammalian species including humans, nonhuman primates, rats, and mice (Bartus and McNaughton, 1980; Birren and Shaie, 1977; Campbell et al., 1980; Doty, 1966; Elias and Elias, 1976; Gold and McGaugh, 1975; Goodrick, 1972; Rigter et al., 1980).

2. REACTIVE SYNAPTOGENESIS IN YOUNG ADULT ANIMALS

The hippocampal dentate gyrus has been employed by our laboratory as a model system to study reactive synaptogenesis (Fig. 1). The dentate gyrus consists of a V-shaped layer of granule cells whose apical dendrites project toward the hippocampal fissure and arborize in an area called the molecular layer. The granule cell axons (mossy fibers) project toward the hilar region of the dentate gyrus and terminate on the CA4 and CA3 pyramidal cells of the hippocampus proper. Our studies have concentrated on the molecular layer of the dentate gyrus

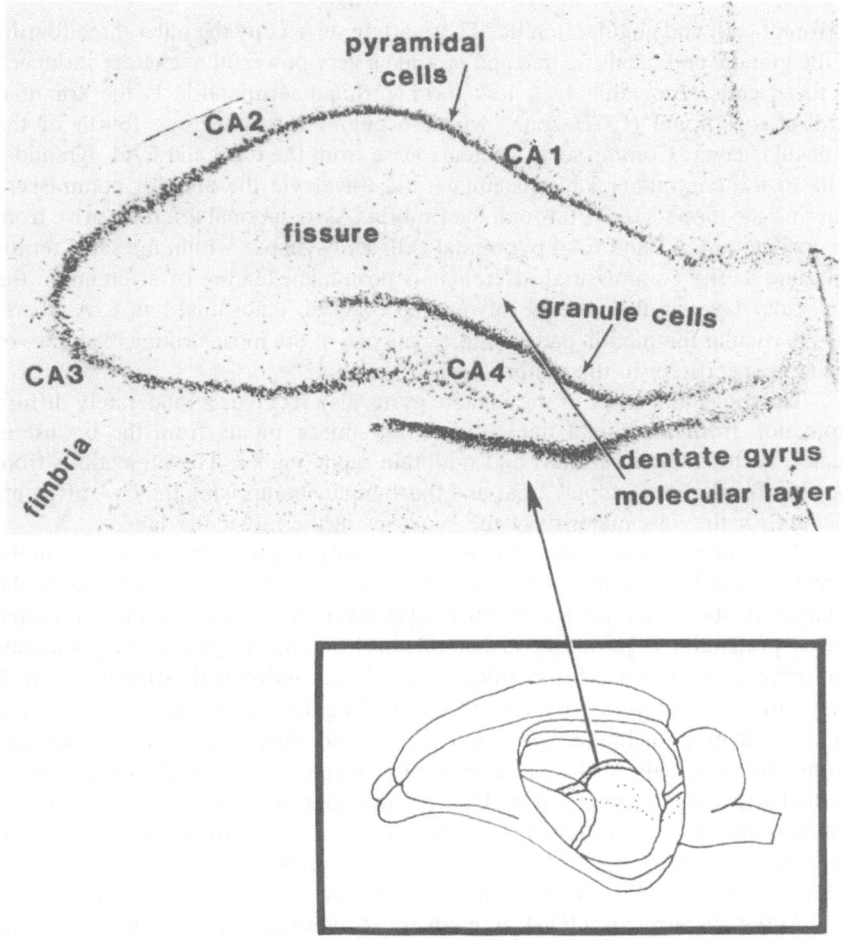

Figure 1. Section cut in a coronal plane through the dorsal hippocampus illustrating the location of the major cell types and the terminology commonly used to describe the different areas of the hippocampus. The hippocampal formation consists of the hippocampus proper, where the pyramidal cells are the major cell type, and the dentate gyrus, where the granule cells are the major cell population. The dendrites of the granule cells arborize in the zone called the molecular layer.

and the subsequent alterations in afferent lamination that occur following partial denervation.

The major inputs to the granule cell dendrites consist of the entorhinal, commissural, and associational afferents. The entorhinal fibers arise primarily from the ipsilateral entorhinal cortex entering the hippocampal formation via the perforant path and angular bundle. These afferents occupy the outer three fourths of the granule cell dendritic tree and provide a very powerful excitatory influence to these cells. Immediately below the entorhinal termination is the commissural–associational (C-A) zone, which occupies the inner one fourth of the molecular layer. Commissural afferents arise from the CA3 and CA4 pyramidal cells in the contralateral hippocampus and travel via the anterior commissure entering the hippocampus through the fimbria. Associational afferents arise from the ipsilateral CA3 and CA4 pyramidal cells and synapse within the same terminal zone as the commissural afferents. A prominent feature of afferents in the molecular layer is the absence of overlap between entorhinal and C-A fibers. This particular lamination pattern represents one of the most striking examples of synaptic specificity in the mature CNS.

The molecular layer of the dentate gyrus also receives a moderately diffuse projection from the septal nuclei, and very minor inputs from the brainstem nuclei, such as locus ceruleus and midbrain raphe nuclei. The projections from the entorhinal cortex, septal area, and those inputs comprising the C-A afferents account for the vast majority of the synapses in the molecular layer.

A technique we have used in the past to study reactive synaptogenesis in the dentate gyrus is to unilaterally remove the entorhinal cortex and to study the changes in the denervated molecular layer over time. Because the entorhinal cortex projection is primarily ipsilateral, each animal's opposite hippocampus can serve as a control. This enables us to rule out individual differences which may introduce variability in the factors which regulate sprouting. With the Fink-Heimer staining technique, the appearance and distribution of degenerating axons and terminals in the outer molecular layer of the dentate gyrus can be studied after an entorhinal lesion. By two days after the lesion, fine degeneration products are evenly distributed throughout the outer three fourths of the molecular layer (Fig. 2). By 30 days postlesion, the appearance of this degenerative debris has been greatly reduced and has become much coarser in appearance.

At the ultrastructural level, immediately following removal of the entorhinal cortex there is a tremendous drop in the density of normal-appearing synapses in the outer three fourths of the molecular layer (Matthews *et al.*, 1976a). The decline in synaptic density is accompanied by a precipitous increase in degenerative debris, characteristic of dying synaptic contacts. Ipsilateral entorhinal input accounts for approximately 85% of the synaptic contacts in this area and nearly 60% of the total synaptic input to the granule cells. Loss of synaptic contacts continues for a period of time but eventually reactive synaptogenesis restores the synaptic density to prelesion levels. The restoration process is initiated within four or five days following the damage and continues for many months. The most reactive portion of the process occurs within the initial 30 days following the perturbation, with a much slower rate occurring afterwards. The reinnervated or restored neuropil looks essentially identical to that in control or normal unoperated animals by 180 days postlesion (Matthews *et al.*, 1976b).

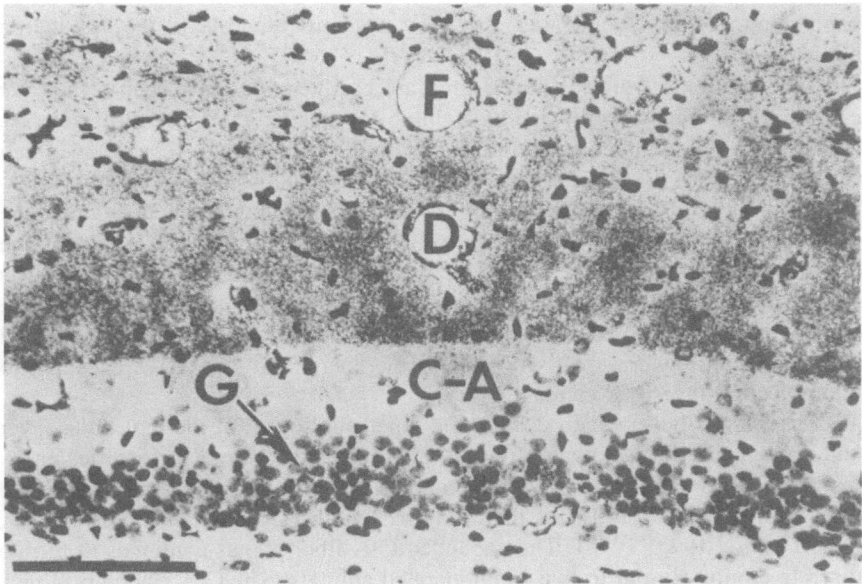

Figure 2. Photomicrograph showing the pattern of degeneration two days after unilateral removal of the entorhinal cortex in a young adult rat. On this coronal section of the dentate gyrus molecular layer stained by the Fink-Heimer method, the boundary between the commissural-associational zone (C-A) and the area denervated by the loss of entorhinal input (D) can be clearly seen. The entorhinal afferents occupy a zone from the hippocampal fissure (F) to the C-A zone. Calibration bar: 100 μm.

A change in the astrocyte population accompanies this reinnervation process. In response to terminal degeneration, astroglia in other CNS tissue have been shown to undergo ultrastructural changes (Laatsch and Cowan, 1967; Bignami and Ralson, 1969; Westrum, 1973). The same is true in the dentate gyrus following a massive denervation. Using Cajal's gold sublimate technique, we can detect a change in the astrocyte population within 48 hr postlesion. Astrocytes appear to hypertrophy and migrate to the area of denervation, resulting in an increase in total number of glia in the denervated neuropil. This situation persists for approximately 7–12 days, and during this time much of the degenerative debris is removed (Rose *et al.*, 1976).

Afferents responsible for this reinnervation have been described in part. The most specific replacement of the lost synaptic contacts would be the regeneration of the entorhinal afferents. However, this does not appear to occur in the adult mammalian CNS. Instead, fibers originating from the contralateral entorhinal cortex appear to proliferate in the denervated zone. This crossed entorhinal projection already exists in a very sparse nature preoperatively, but following a lesion it becomes much more extensive and is responsible for a major portion of the replacement (Steward *et al.*, 1974). Afferents originating from the diagonal band region of the septal area also play a significant role in the restoration process.

The afferents from the septal area are easily monitored because they are cholinergic and can be followed by staining for the enzyme acetylcholinesterase (AChE) (Fig. 3). In the normal brain, and on the contralateral side following an entrohinal lesion, the dentate gyrus molecular layer shows a light, evenly deposited AChE stain. Most of the staining is located in a zone immediately above the granule cells. In marked contrast to this staining configuration is that of the dentate gyrus ipsilateral to the lesion. A very dark, intense staining deposit is observed in the outer molecular layer, indicative of the proliferation of the septohippocampal projection. Direct biochemical measurements on AChE and choline acetyltransferase activity support the contention that the septal input to the molecular layer has increased.

The C-A fibers, which occupy the inner one fourth of the molecular layer and are outside the denervated region, also respond to the entorhinal lesion by expanding outward into the denervated area. These fibers increase their terminal field from 130% to 140% of their original width (Lynch *et al.*, 1976, 1977; Scheff *et al.*, 1977, 1978). The expanded C-A fiber plexus continues to form a discrete lamina similar to that in unoperated animals. The C-A fibers form new synapses in their newly acquired territory, actually invading part of the former zone of the ipsilateral entorhinal input (Fig. 4).

This fiber outgrowth following a loss of the entorhinal input proceeds at a specific rate and the magnitude of the C-A growth is quantifiable. A number of techniques have been used to study this outgrowth and provide substantial documentation for sprouting of the system following a lesion of the entorhinal cortex (Goldowitz and Cotman, 1980; Lynch *et al.*, 1976; Scheff *et al.*, 1977; West *et al.*, 1975). Furthermore, neurophysiological studies have also demonstrated that the C-A system has not only sprouted but formed new synapses that are functional (West *et al.*, 1975).

3. NORMAL ANATOMY OF THE AGED HIPPOCAMPUS

3.1. Light Microscopic Analysis

Before studying the capacity of the aged brain to respond to partial denervation, it was necessary to examine the normal anatomy of the hippocampus to assess any anatomical changes, if any, which might occur solely as a result of the aging process. A group of aged (24- to 28-month-old) Sprague-Dawley rats were sacrificed and the brains processed. Analysis of the hippocampus revealed that all major cytoarchitectonic fields could be identified. While no detailed cell counts were made of the pyramidal and granule cells in the hippocampus, it was determined that all areas appeared to have a full complement of cells. Other

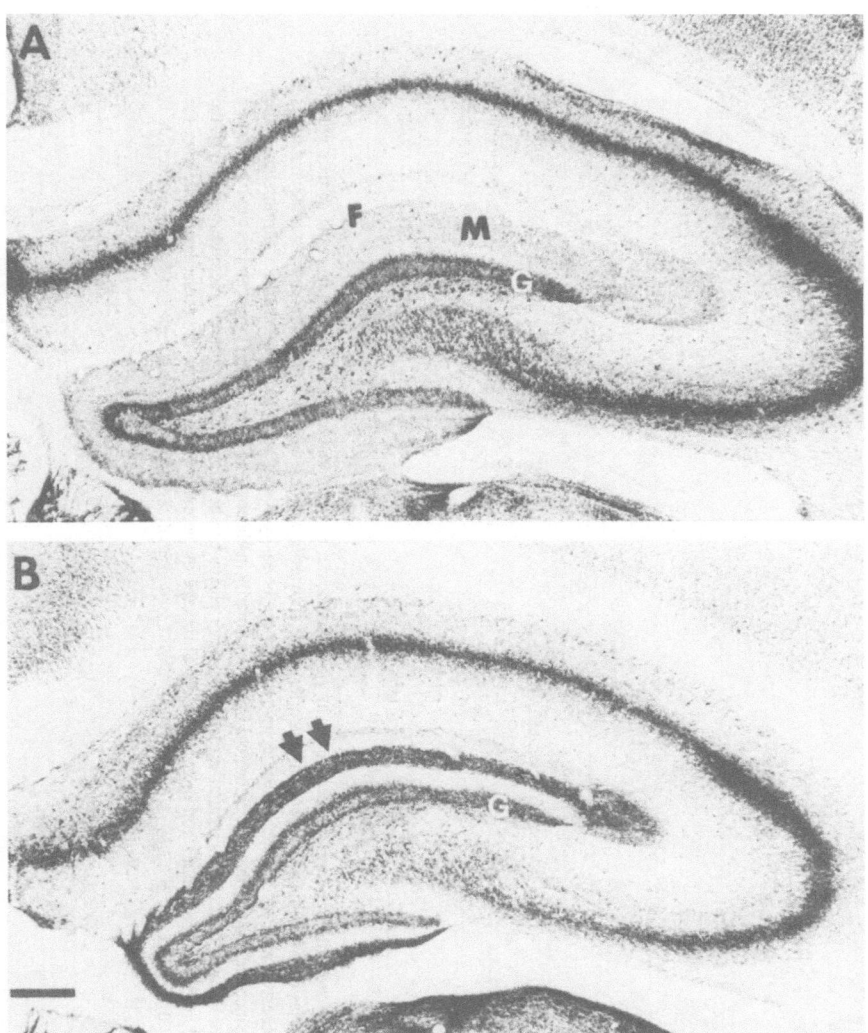

Figure 3. Coronal sections of the dorsal hippocampus stained for the enzyme acetylcholinesterase. (A) Photomicrograph from an unoperated animal showing an even distribution of staining in the molecular layer (M) extending from the granule cell layer (G) to the fissure (F). (B) Photomicrograph showing the change in staining 15 days after ipsilateral removal of the entorhinal cortex. The arrows indicate the intense band of staining in the outer portion of the molecular layer. Immediately above the granule cell layer (G) there appears a wide clear zone. Calibration bar: 300 μm.

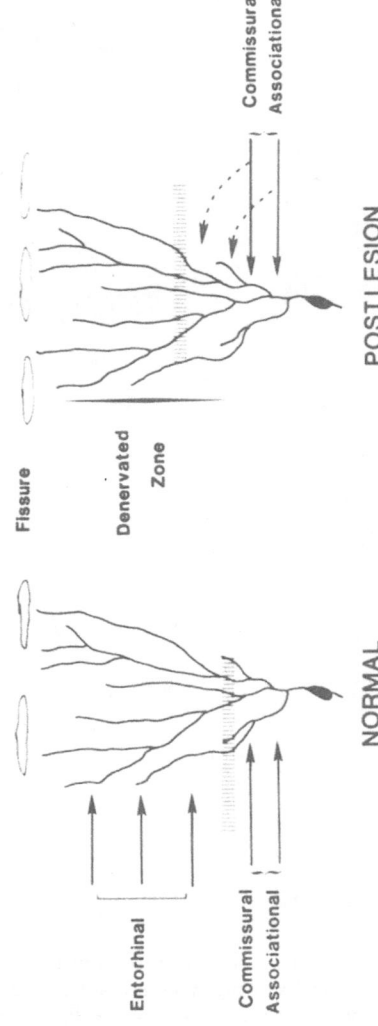

Figure 4. Schematic drawing of the changes in the commissural-associational afferents remaining in the dentate gyrus following a lesion of the ipsilateral entorhinal cortex. The outward movement of the commissural and associational fibers has been observed with several anatomical techniques and can be easily quantified by drawing the area of termination of the afferents. Since the projection of the entorhinal fibers is primarily ipsilateral, the contralateral hippocampus can be used as a control. The growth of the commissural-associational fiber plexus can be expressed as a percent increase in width of the denervated side over the contralateral or control side. Shrinkage of the outer portion of the molecular layer following the entorhinal ablation is not a factor since only the absolute width of the commissural-associational fiber projection is measured.

laboratories studying the hippocampus of 33-month-old rats observed no changes in total volume of the hippocampus, in total number of granule and pyramidal cells, and also no change in the terminal field of the hippocampus mossy fibers as compared with younger controls (P. Coleman, personal communications, 1983). Our own measurements of the different strata within the hippocampus and dentate gyrus revealed no obvious significant difference between the two age groups.

Subsequent sections were stained for the enzyme AChE, and an analysis was made of the distribution of staining. The most prominent features were a densely stained band which lies just above the granule cell layer, a moderately stained area in the outer three fourths of the molecular layer, and a somewhat paler staining in the C-A area. The CA1 region showed a much paler staining in stratum radiatum and extremely light staining in stratum lacunosum-moleculare of regio superior. This composite of AChE staining in the hippocampus was identical to that found in the younger animals and indicates that the septohippocampal circuitry was intact in these aged rats. Such a conclusion is supported by previous biochemical measurements on the hippocampus which show that acetylcholine activity and AChE levels in both young adult and aged rats are very similar (Cotman and Scheff, 1979).

Sections were also stained with the Holmes's fiber stain to assess the entire fiber plexus of the molecular layer. There appeared to be no discontinuities in the dentate gyrus and both young and aged animals revealed a very dense fiber milieu. Thus, commissural, associational, and entorhinal afferents to this region appeared equivalent for the two age groups.

Finally, sections were stained with the Cajal gold sublimate technique for astrocytes. Sections of the hippocampus from the brains of both age groups, taken within 2 mm of the level at which the hippocampal arch begins its ventral descent, were evaluated. A standardized area within the dentate molecular layer was assessed for the total number of astrocytes stained. In agreement with previous studies, the astroglia of each animal appeared randomly distributed throughout most of the sections. The most significant finding was an increase in total number of astrocytes with age (Table 1). While no astrocytic plaques appeared in this area, the astrocytes of the aged animals did appear hypertrophied in agreement with other reports (Landfield et al., 1977). The processes of these cells were very thick and coarse in appearance and the cell bodies appeared to be much larger and more swollen than those observed in younger animals (Scheff et al., 1980).

Age-related alterations in cerebral blood flow and vascular structures have been described in human brain and we thought it might be important to analyze changes in the senescent rat brain. We felt that changes in the microvasculature may be of importance in the maintenance of brain metabolism and hence in recovery from brain damage. Animals from both age groups were anesthetized

Table 1. Number of Astrocytes Counted in a Standard Area in the Outer Molecular Layer of the Denate Gyrus in Unoperated Animals at Different Ages[a]

2 Months old	3 Months old	12–18 Months old	25–30 Months old
53	68	67	78
62	56	68	92
51	52	71	90
53	61	64	108
56	63	80	87
64	50	72	89
48	53		
50	50		
54	62		
$\bar{X}$ 54.5	57.22	70.3	90.6
SD ± 5.34	± 6.49	± 5.53	± 9.79
N 9	9	6	6

[a]Each value in a column represents the mean value for a single animal at that age. Counts were made within a grid 650 × 150 μm for each section. Values for each subject are means obtained from counting six sections per animal. [From Scheff et al. (1980).]

and perfused with India ink. Frozen sections were then cut and laminar and vascular measurements were taken from the molecular layer of the dentate gyrus (Fig. 5). Total vessel length within a defined and standardized area was determined from camera lucida drawings of the vascular bed. The aged animals showed a significant decrease in capillary bed density as evidenced by a 17.5% drop in total capillary length in the molecular layer. There appeared to be no change in the width of the molecular layer or any significant change in the diameter of the capillaries as measured from semithin plastic sections. The most notable alteration in the appearance of the vascular supply was a change from the delicate pattern of the vessels normally seen in young adult animals. The aged animals revealed vessels which appeared much coarser in nature and seemed to have a more tortuous route. The pattern of vascular architecture of the aged animals appeared to be less complex than that of the young adult rats. This significant decrement in capillary density and appearance may contribute significantly to metabolic alterations observed in senescence.

3.2. Electron Microscopic Analysis

The electron microscope has been an extremely valuable tool for detecting age-related changes within the neuropil of the hippocampus and other brain areas. Circumspect observation of electron micrographs taken from aged rodent hippocampus reveals no obvious changes in morphology of the dendrites, axons, or synaptic contacts. Hasan and Glees (1973) reported that the hippocampus of

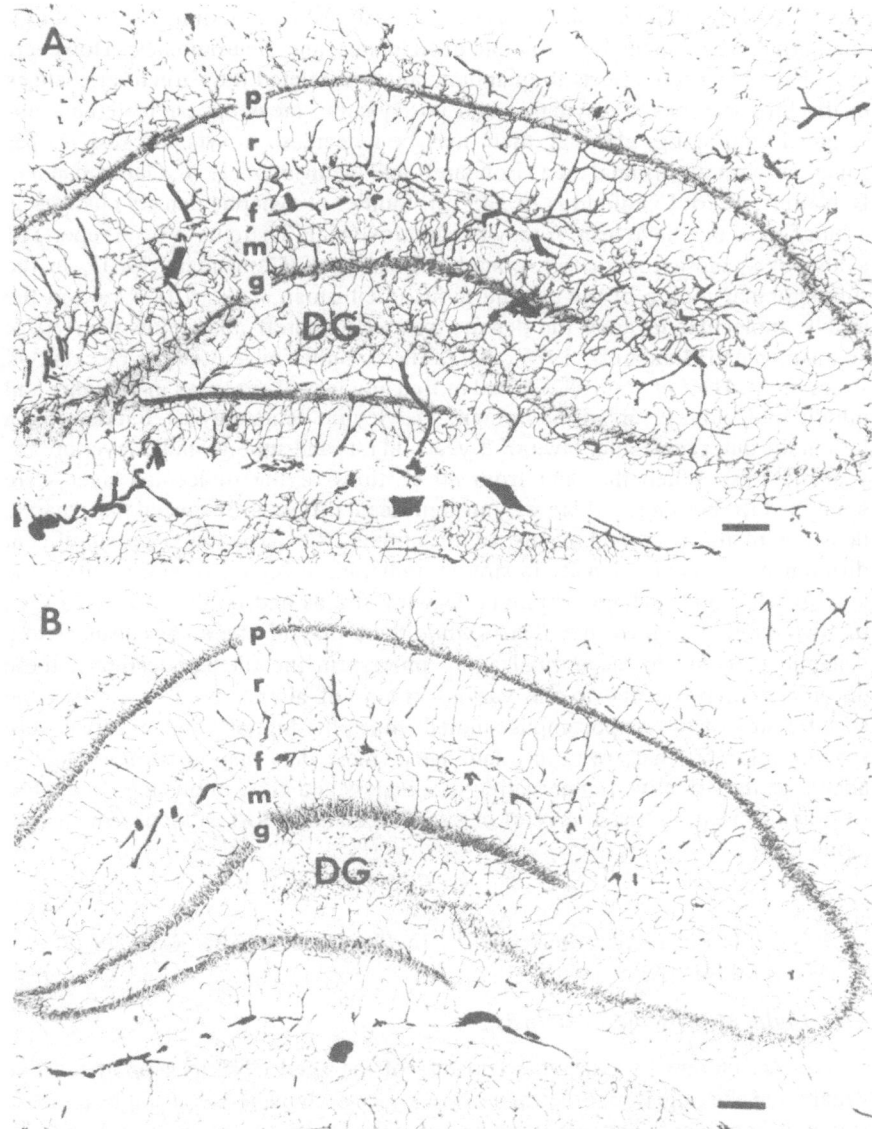

Figure 5. Coronal sections taken from the dorsal hippocampus of animals perfused with black India ink to demonstrate the vascular supply to this area. The sections were subsequently stained with cresyl violet to enhance cellular boundaries. The section taken from the aged animal (B) shows a noticeable decrease in vessel number and complexity particularly when compared with the younger animal (A). p, Pyramidal cell; r, stratum radiatum; f, fissure; m, molecular layer; g, granule cell; DG, dentate gyrus. Calibration bar: 300 μm. [From Cotman and Scheff (1979).]

aged rats seems to have a higher electron density of axon terminals as well as a qualitative increase in dendritic neurofilaments and neurotubules. However, there have been few efforts to quantitate any age-related ultrastructural changes in the hippocampus. Presynaptic and postsynaptic areas appear to remain unchanged with aging, although there may be a decrease in synaptic vesicles per square micron of terminal area (Bondareff and Geinisman, 1976; Geinisman and Bondareff, 1976; Landfield *et al.*, 1979). Studies of synaptic density changes as a function of age have been somewhat controversial. There have been reports of a loss of synapses contacting granule cell somata (Geinisman, 1981), in the supragranular region of the dentate gyrus (Geinisman *et al.*, 1977), and in the middle molecular layer of the dentate gyrus (Bondareff and Geinisman, 1976; Geinisman and Bondareff, 1976). All of these studies have used the Fischer 344 strain of rat. Hoff *et al.* (1982) failed to find any loss of synapses in the inner and middle molecular layers of the Sprague-Dawley rat dentate gyrus, but detected a loss in the outer molecular layer of 2-year-old rats and supragranular layer of 2.5-year-old rats. When the data from all portions of the molecular layer were combined for both ages, there was no significant difference between the synaptic densities of the aged and young adult animals. We have found no significant difference in synaptic density in stratum radiatum of regio superior in either the aged Fischer 344 and aged Sprague-Dawley strains and no difference between the two strains at either age. Thus, while there may not be a consensus as to a synaptic loss during the normal aging process in the rat hippocampus, these differences may be due to the various strains of animal used as well as the specific area of the hippocampus studied. Since the hippocampus receives very specific afferents that are rigidly laminated, these data may prove important in analyzing the effective reinnervation of a denervated zone in the aged animal as well as determining the source(s) of synaptic contacts lost as a consequence of aging.

4. REACTIVE SYNAPTOGENESIS IN THE DENTATE GYRUS OF AGED ANIMALS

4.1. *Light Microscopic Analysis*

In an attempt to assess the capacity of the aged CNS to support reactive synaptogenesis, aged male Sprague-Dawley animals were subjected to an entorhinal lesion identical to that previously employed for young adult animals (Scheff *et al.*, 1980). At various intervals following entorhinal removal, animals were examined to assess the reaction of the septal and C-A afferents. Initially, the hippocampi were stained with a modified Fink-Heimer stain two days after the ablation. The pattern of degeneration products in aged animals were identical to that found in younger animals. This supported our previous conclusion that the pattern of entorhinal input to the dentate gyrus does not change with age.

Brains were also stained for the enzyme AChE to assess reactive growth of the septohippocampal afferents. Young adult animals killed at 12 days postlesion were characterized by a wide dark band of AChE staining in the outer half of the dentate molecular layer ipsilateral to the lesion. The darkly stained supragranular band was not altered, but the inner portion of the molecular layer, the C-A zone, showed a clearing or absence of staining deposit. Aged animals at 12 days postlesion appeared qualitatively different from that of younger animals. Aged rats also featured darkening of the band in the outer portion of the molecular layer, but the band was not as intense as that observed in the younger animals. As in the young animals, the supragranular band was not altered and a clearing of the C-A zone was evident. Thus, the pattern of the AChE staining was identical for the two age groups and only the intensity of staining in the outer three fourths of the molecular layer appeared different. These results cannot be explained on the basis of difference in molecular layer volume between the two age groups. Although the molecular layer shrinks in both aged and young rats following an entorhinal lesion, measurements failed to reveal a significant difference in width of the molecular layer between young and aged animals.

Sections were stained with the Holmes method to reveal changes in the C-A fibers. Removal of the entorhinal cortex in the young 3-month-old animals resulted in a 21.7% increase in the width of the C-A fiber plexus at 12 days postlesion. The expansion in the older animals was significantly lower, showing only a 13.4% increase at 12 days following an identical denervation (Fig. 6). Considerable variation occurred in the aged animal group. In some cases the response was nearly identical to that of the younger rats while other aged animals demonstrated practically no outgrowth of the fiber plexus. This variability was not correlated with any obvious abnormal feature of the dentate gyrus anatomy, nor to the size and location of the entorhinal lesion.

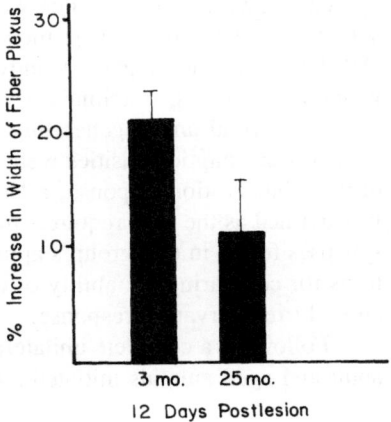

Figure 6. Bar graph showing the change in outgrowth of the commissural-associational fiber plexus as a function of age. At 12 days postlesion the two groups differed significantly ($P < 0.01$), indicating a decreased growth response in the aged animals.

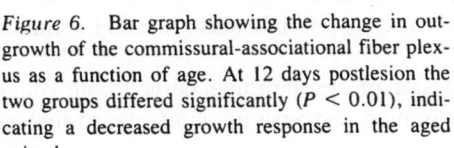

The response of astroglia was also investigated following the entorhinal ablations in both aged and young rats. As previously described, glia of the 3-month-old animals appeared to migrate outward from the inner one third of the molecular layer and form a band of cells running parallel to the granule cell layer. This migration of astrocytes coupled with shrinkage in the molecular layer resulted in an increase in the density of astrocytes in the outer three fourths of the molecular layer. At twelve days postlesion the astroglia in the younger animals appeared morphologically identical to those from control animals. Astrocytes at 12 days postlesion in the aged rat had swollen cell bodies and processes which were extremely large and coarse. The astrocytes had a morphologic appearance which suggested maximal activity. This was not surprising since, as mentioned previously, the astroglia in aged animals are markedly hypertrophied even in the unoperated animal. As in the young animals there was a significant increase in the number of astrocytes occupying the outer molecular zone at 12 days postlesion.

Many features of the hippocampal neuropil are indistinguishable or at best marginally distinguishable between mature and aged animals. However, the response of residual afferents in this area is markedly dependent on age. Reactive growth is apparently not as extensive in aged animals. Still, a number of important questions remain unanswered. Is reactive growth still occurring beyond the time limits of these studies? Do the rates of reinnervation differ between the two age groups? These data do not indicate whether the final magnitude of the plasticity is actually reduced, or whether the aged animals' response is simply occurring at a slower rate.

4.2. Ultrastructural Analysis

In this study we used quantitative electron microscopy to examine responses within the denervated outer two thirds of the dentate gyrus molecular layer (Hoff et al., 1982). Both young adult (3-month-old) and aged (26- to 30-month-old) Sprague-Dawley male rats underwent a complete unilateral lesion of the entorhinal cortex. The animals were then allowed to survive for 2, 4, 10, 60, 120, or 180 days after the surgery. Control data were obtained from both unoperated young adult and aged animals.

All normal and degenerating synaptic profiles as well as vacant or unoccupied postsynaptic densities were counted. To provide an objective evaluation of the reinnervation response, a "reinnervation index," or RI_{50} was used. This was defined as the time required after a lesion for the replacement of 50% of the synapses found in that group's appropriate control animal. The RI_{50} served as a basis for comparing the ability of experimental animals to initiate and maintain an early reinnervation response.

Following a complete unilateral lesion of the entorhinal cortex, both young adult and aged animals initiated a restoration response in an attempt to replace the lost synaptic contacts. The major difference between the young and aged animals was that the reinnervation began very quickly for the young adult animals and was significantly ahead of the aged animals by ten days postlesion. At this point, the young animals showed a 28% restoration of the neuropil while the aged animals showed only 18% replacement. At 60 days, synapse replacement in young adult animals was still significantly advanced as compared with the aged rats, but by 180 days both young and aged animals had attained similar synaptic densities (Fig. 7). This delay in the reinnervation process observed in the aged animal group was further shown by comparison of the RI_{50}. The aged animals required, on an average, ten days longer to replace 50% of the synapses in the outer two thirds of the molecular layer then the young adult animals.

Several factors may have restricted the reinnervation process: (1) appearance of degeneration, (2) clearance of the degeneration products, (3) growth of the reactive afferents, or (4) assembly of new synaptic contacts. Degeneration and the clearing of the degeneration appears to precede the reinnervation of the denervated zone. In many cases astrocytes may play an essential role in this process. Light microscopic studies indicate that astrocytes in unoperated aged animals appear hypertrophied in the molecular layer of the dentate gyrus (Landfield *et al.*, 1977). This alteration in the glial cells may be related to the change in the reinnervation process.

Electron-dense and electron-lucent degeneration were found to be present throughout the denervated outer two thirds of the molecular layer. Dark degener-

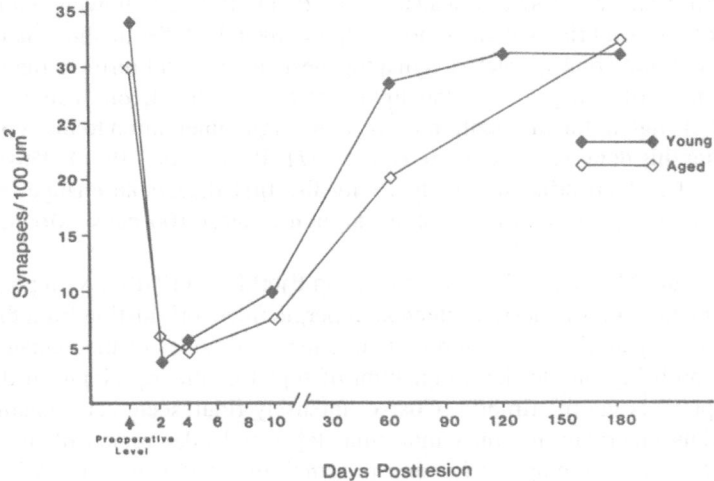

Figure 7. Time course for synapse replacement in the denervated zone of the molecular layer of the dentate gyrus after an entorhinal lesion in young adult (♦) compared with aged rats (◇). By 180 days postlesion both groups have attained equal synaptic density.

ation reached a maximum in both age groups by two to four days postlesion. The animals demonstrated a biphasic response in clearing of the degeneration. In the initial phase about 45% of the debris was removed by ten days postlesion while the remaining degeneration products were removed more slowly. By 60 days after the damage, only 5% of the degenerating terminals remained. The aged animals followed this same pattern but were delayed in the initiation of clearing response. A clearance index (CI_{50}) similar to that used for reinnervation (RI_{50}) indicated that once the removal began in the aged animals it occurred at approximately the same rate as in the younger animals. This index also showed that the aged animals required 2.3 times longer to remove 50% of the dark degeneration than young adults. The important age-related feature of the degeneration process is that aged animals lacked the ability to initiate the early rapid phase of the degeneration removal which was seen in the young adult animals.

5. ADRENERGIC SPROUTING IN THE LIMBIC SYSTEM OF AGED RATS

Our laboratory (Scheff et al., 1978) wanted to know whether the aging mammalian brain was capable of supporting axonal growth in areas other than the hippocampus. In order to answer this question we chose to study the septal nuclear complex. Raisman and co-workers have previously shown that this area undergoes a dramatic sprouting response following partial denervation (Field et al., 1980; Raisman, 1969a,b; Raisman and Field, 1973). Following a unilateral lesion of the fimbria, which removes approximately 50% of the synapses on septal dendritic spines, reactive synaptogenesis begins and restores the original complement of synapses. At the light microscopic level, one can observe a marked change in the distribution of the adrenergic innervation to the septal area following this denervation (Moore et al., 1971; Reiss et al., 1978). This growth response has a specific time course with the first detectable change occuring about 15 days postlesion and lasting through at least 100 days (Moore et al., 1974).

Loy and Moore (1977) and Stenevi and Bjorklund (1978) have reported that following this fimbrial lesion catecholaminergic fibers originating from the superior cervical ganglion grow into certain denervated zones of the dentate gyrus. This is probably due to the elimination of septal cholinergic input to this area which projects via the fimbria. Coarse, intensely flourescent catecholaminergic fibers thus innervate the inner molecular layer and hilar region of the dentate gyrus (Fig. 8). We employed young (3-month-old) and aged (26- to 30-month old) Sprague-Dawley rats and transected the fimbria-fornix unilaterally at a point between the septum and hippocampus. Control animals at both ages demonstrated equivalent flourescence in the medial and lateral portions of the septum. The only preoperative difference between the two groups was the presence of

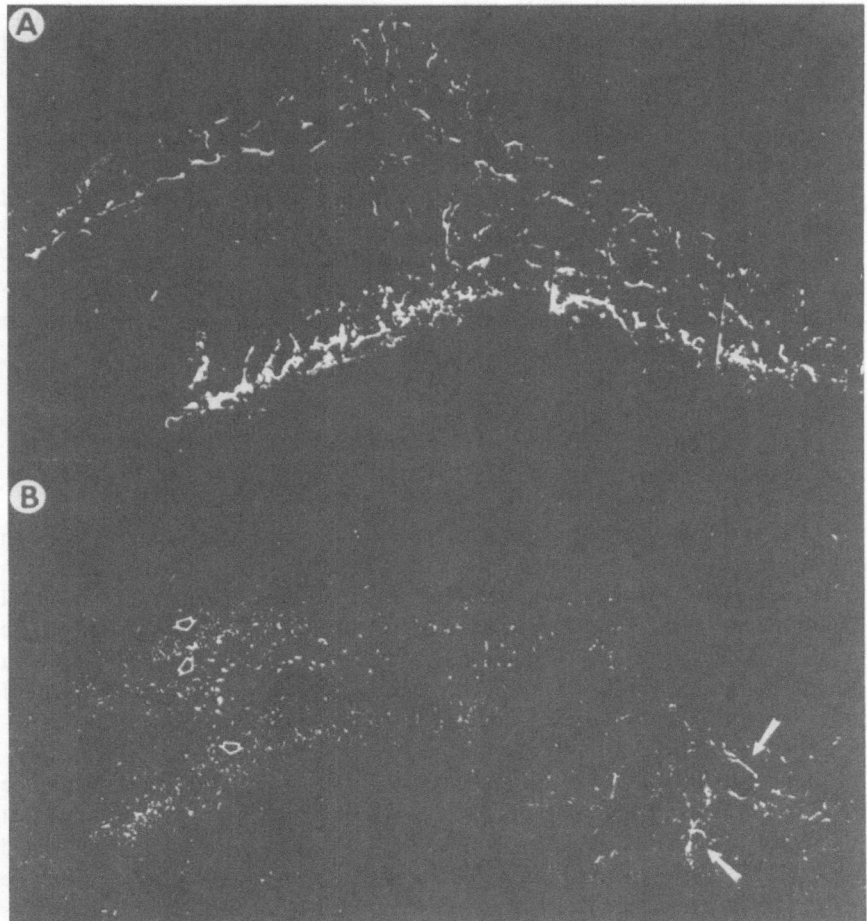

Figure 8. A photomontage of the dentate gyrus 60 days after a unilateral transection of the fimbria showing the anomalous growth of the superior cervical ganglion into the dentate gyrus ipsilateral to the lesion. The response in 3-month-old animals (A) is well defined, localized within the granule cell layer and hilar region. The response of the 26-month-old animal (B) is markedly reduced (solid arrows). The remaining flourescent elements (open arrows) in the aged animal are autofluorescent granules that are routinely observed in aged animals. [From Scheff *et al.* (1978).]

yellow-brown autofluorescent material found in many areas of the CNS of the aged animals. By 30 days after the fimbria transection in the younger animals, the catecholaminergic (CA) fibers had proliferated in the denervated septal area in agreement with previous reports (Moore *et al.*, 1971, 1974; Reiss *et al.*, 1978). The same operation performed on the older animals elicited a qualitatively similar but quantitatively less pronounced response in the septal area. These results indicated that the fibers still have the capacity to grow in the aged brain but to a reduced degree.

The reaction of the sympathetic neurons in the hippocampus following fimbrial transection was then monitored. In young adult animals coarse, intensely flourescent CA fibers (characteristic of sympathetic origin) innervated the dentate gyrus molecular layer and the hilar region after the lesion. In the aged animals this anomalous growth response was either totally absent or greatly reduced as compared with the younger animals. These experiments, involving the sympathetic innervation of the dentate gyrus and reactive growth in the septum, indicate that the reduction in plasticity found with age severely effects the central and peripheral nervous systems.

6. POSSIBLE MECHANISMS FOR REDUCED SPROUTING CAPACITY

The previously presented experiments demonstrate that aged animals can support reactive synaptogenesis but do so in a reduced capacity. This reduction may result from a variety of factors, e.g., a significant change in the neuropil environment, an increased threshold to a growth stimulus, or an inability of the target cells to accept new synaptic contacts. There may simply be a lack of a growth-promoting substance or an increase in a growth retardant. Many or all of these factors may play important roles in aged animals. We have to date approached one aspect of the problem.

One possible explanation for the reduced growth in aged animals could be the direct or indirect action of glucocorticoids. Several studies have reported that glucocorticoids are elevated in aged rats (Landfield et al., 1978; Riegle and Hess, 1972; Tang and Phillips, 1978) and that the control of corticosterone (CORT) levels by the pituitary adrenal axis is altered with aging (Hess and Riegle, 1970; Landfield et al., 1978). We subsequently determined the serum CORT levels, using a radioimmunoassay (RIA), at three points in the diurnal cycle: 0800, 1800, and 23.30 hr in animals at 3–6 months of age (young), 14–17 months (middle), and 26–28 months (old). At all three time points in the diurnal cycle, old animals had significantly higher mean levels of CORT than young animals (Table 2). This change was seen initially in the middle-aged animals, where peak CORT levels at 1800 hr were above that found in the young rats. The 1800 hr levels of CORT were 24.0 μg/100 ml serum for the middle-age and aged animals while those for the young animals were 14.9 μg/100 ml serum. These values were the peak values for each group respectively. The 23.30 hr rise was the most dramatic in old animals, and suggested that significantly higher circulating levels of CORT were present in the senescent rat. Furthermore, these high levels in the oldest group persisted for a greater percentage of the diurnal cycle.

Table 2. Tabulated Means for Serum Corticosterone

Age	0800 hr	1800 hr	2400 hr
Young	2.12 ± .79	14.91 ± 1.79	9.7 ± 2.56
(3–6 month)	(N = 10)	(N = 10)	(N = 8)
Middle age	1.2 ± .28	24.31 ± 2.65[b]	13.47 ± 3.48
(14–17 months)	(N = 10)	(N = 10)	(N = 8)
Old	3.49 ± .94[a]	23.57 ± 2.63[c]	19.56 ± 3.49[d]
(25–28 months)	(N = 10)	(N = 9)	(N = 8)

[a] $P < 0.05$ compared with middle aged animals at same time of day.
[b] $P < 0.01$ compared with young animals at same time of day.
[c] $P < 0.02$ compared with young animals at same time of day.
[d] $P < 0.05$ compared with young animals at same time of day.

Since the circulating levels of CORT were significantly higher in aged animals and previous data indicate a decrease in the sprouting response with age, we sought to examine this relationship further using young adult animals. Sprague-Dawley rats were bilaterally adrenalectomized and allowed ten days of convalescence before the start of the experimentation. Five days prior to an entorhinal ablation, a pellet containing a specified concentration of CORT was subcutaneously implanted in each animal's neck. By varying the ratio of cholesterol and CORT it was possible to control the hormone dosage. These pellets in conjunction with the adrenalectomies allow for a precise regulation of the circulating concentrations of glucocorticoid in each animal.

The serum concentration of CORT was maintained at one of four different serum levels: 3.5 μg/100 ml, 14.0 μg/100 ml, 28 μg/100 ml, and less than 0.1 μg/100 ml. This last group was given cholesterol pellets in place of CORT. In addition, a control group which received a sham adrenalectomy and no steroid treatment (Sham) was included. Five days following implantation of the pellets the animals were subjected to a unilateral removal of the entorhinal cortex.

Following a fifteen-day survival time, the animals were killed and the brains analyzed for changes in brain morphology indicative of axon sprouting. The different levels of serum CORT had a significant influence on the outgrowth of the C-A afferents in the dentate gyrus. The results are summarized in Fig. 9. Adrenalectomized animals given only cholesterol and no CORT (Choles) did not differ from the Sham control animals. In addition, animals maintained at the 14.0 μg/100 ml serum level also did not differ from the Choles group or the Sham control animals. This was not suprising since the peak circulating levels of CORT in young animals is approximately 14.0 μg/100 ml serum. However, animals maintained at the peak levels observed in the aged animals, 28 μg/100 ml serum, had a sprouting response significantly lower than the other groups, supporting our hypothesis that elevated glucocorticoid levels in aged animals may be one of the factors that inhibit the reactive growth response.

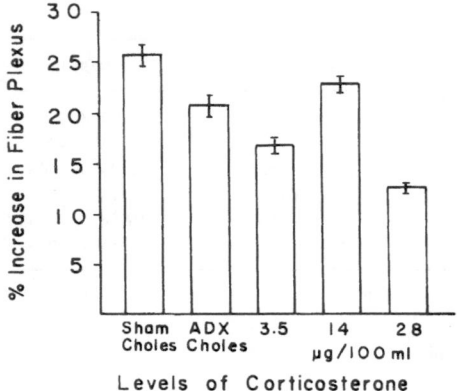

Figure 9. Bar graph showing changes in the commissural-associational fiber plexus 15 days after unilateral removal of the entorhinal cortex in young adult adrenalectomized (ADX) animals. Corticosterone concentrations are expressed in µg/100 ml serum. Control animals were administered cholesterol (Choles). Animals maintained at extremely high concentrations of corticosterone showed a significant reduction in fiber outgrowth. Bars show mean ± SE. [From Scheff *et al.* (1982).]

As previously mentioned, the astrocyte population in aged animals is quantitatively and qualitatively different from that found in young adult animals. Thus we wanted to know if the different levels of CORT also had an affect on the astrocytes. Again, animals were adrenalectomized and received CORT therapy matching the concentrations in each of the groups above. The animals used for the astrocyte analysis were not subjected to an entorhinal ablation. Twenty days after the pellet implantation, the animals were killed and the brains stained with the Cajal gold sublimate stain. Counts were made of the number of astrocytes in the outer molecular layer of the dentate gyrus. In all of the animals the astroglia appeared to be randomly distributed throughout the entire hippocampus. In animals maintained at the extremely high levels of CORT (28 µg/100 ml serum), the astroglia appeared hypertrophied. Their cell bodies were irregular in shape and had processes which appeared thicker than those in naive animals. This was not the case in animals maintained at lower levels of CORT. The 28 µg/100 ml CORT animals had significantly more cells than the controls and in fact more than any other group. Their levels, however, compared very well with that previously observed with the aged animals.

In an effort to ascertain whether or not steroid therapy disrupts reconstruction of the denervated neuropil, ultrastructural analysis was also employed. While previous light microscopic analysis has shown a reduction in the growth of specific afferents, it has only been assumed that a reduction in synapse formation also occurs. It is possible with ultrastructural analysis to monitor not only the magnitude but also the rate of synapse replacement.

In these experiments, young adult male Sprague-Dawley rats were used and subjected to a bilateral adrenalectomy and maintained on two different glucocorticoid regimes, 28 µg/100 ml serum and less than 0.1 µg/100 ml. A control group was included which was sham adrenalectomized and given cholesterol.

The animals were then subjected to a unilateral entorhinal lesion and killed at 4, 10, 30, and 60 days postlesion. Montages were constructed of the molecular layer of the dentate gyrus dorsal leaf ipsilateral to the entorhinal ablation. Every synaptic profile was counted and assigned as either normal or degenerating.

In the outer three quarters of the molecular layer ipsilateral to the lesion, there was approximately an 82% decrease in the number of normal synapses for both the control and adrenalectomized animals given cholesterol (ADX) at four days postlesion (Fig. 10). This is in agreement with previous results for the dorsal leaf of the dentate gyrus in young adult animals (Hoff *et al.*, 1982). In contrast, animals adrenalectomized and treated with CORT (ADX CORT) showed only a 54% decrease. At ten days following the entorhinal lesion, both the control group and ADX animals showed some synpatic replacement. These groups demonstrated a 125% increase in normal-appearing synapses as compared with the value observed at day 4 postlesion. The ADX-CORT-treated animals showed an additional 61% decrease, thus indicating that terminal death occurred at a reduced rate or at least was delayed in this group (Fig. 11). By 60 days postlesion the control and ADX animals had significantly replaced many of the lost contacts and were found to be within 75% of the preoperative levels. ADX-CORT-treated animals are substantially retarded in synaptic replacement, showing only 40% of the preoperative synaptic density. This study supports the light

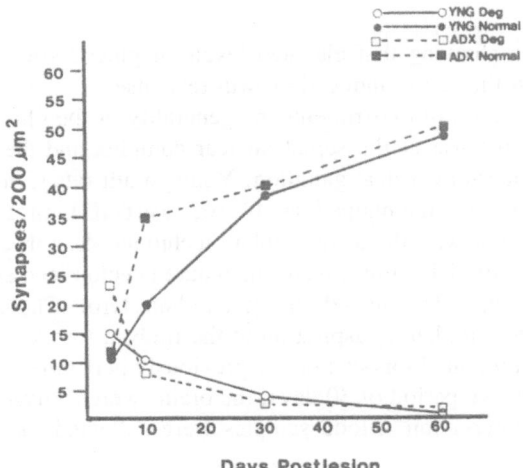

Figure 10. Time course for the removal of degeneration products and the reappearance of normal synapse density in the molecular layer of the dentate gyrus after an entorhinal lesion in young adult animals (YNG) and animals given an adrenalectomy (ADX) prior to the entorhinal ablation. Note that by 30 days postlesion both groups have the same synaptic density. The removal of the degenerative debris proceeds at the same rate in both groups.

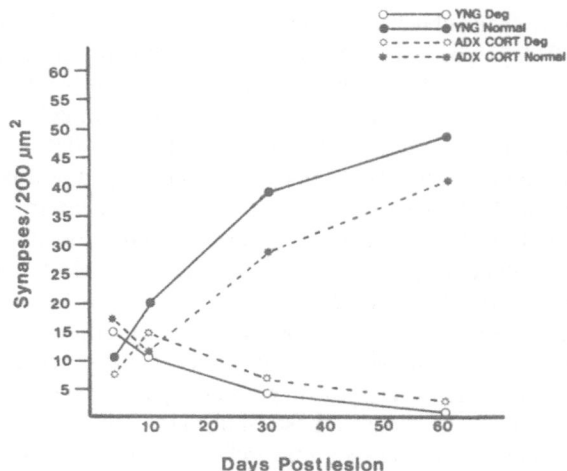

Figure 11. Time course for the removal of degeneration products and the reappearance of normal synaptic density in the molecular layer of the dentate gyrus after an entorhinal lesion in young adult animals (YNG) and animals adrenalectomized and maintained on high levels of corticosterone (ADX CORT). Note that the animals maintained on corticosterone have a significantly lower density of normal synapses at 10, 30, and 60 days postlesion as compared with young adult animals. The young adult animals show a more rapid loss of synapses suggesting that the corticosterone treatment helps preserve the neuropil.

microscopic data showing that elevated levels of glucocorticoids may play an important role in the lesion-induced growth response.

In the next series of experiments the generality of the glucocorticoid effect was tested with the use of the septal nuclear complex and the response of the growth of the superior cervical ganglion. Young adult rats (3 months old) were adrenalectomized and maintained at 14 μg CORT/100 ml serum or 28 μg CORT/100 ml serum with the same implant technique described in the previous experiments. Additional animals were sham adrenalectomized and given cholesterol (Sham) or adrenalectomized and given cholesterol (Choles). The animals were subjected to a unilateral aspiration of the fimbria-fornix at a point midway between the septum and hippocampus as previously performed on aged animals. After a postoperative period of 30 days, the brains were analyzed for changes in catecholamine innervation. Blood samples were collected for RIA analysis of CORT levels.

In all animals maintained at 28 μg CORT/100 ml serum, there was a significant reduction in sprouting of the CA fibers in the septum as compared with sham-adrenalectomized animals and animals maintained at 14 μg CORT/100 ml serum. These latter two groups did not differ in sprouting of the CA fibers in the septum.

In these same animals we measured the anomalous growth of the superior cervical ganglion fibers as previously mentioned. To quantify the sprouting response the hippocampal dentate gyrus region on each side of the brain was visually divided into four different anatomical zones. Zone 1 occupied the most medial aspect of the dentate gyrus; zone 4 occupied the most lateral aspect (which included the pyramidal cells of the CA3a hippocampal region). Since the growth response proceeded in a medial-to-lateral direction, it was possible to assign a quantitative value to the growth response by determining the most lateral zone into which the fibers had grown. At least six consecutive sections were assessed for extent of sympathetic outgrowth and a mean was obtained for each animal.

Control and ADX animals given cholesterol were indistinguishable in their response. Young adult animals maintained at 14 μg CORT had fibers which appeared finer in character than those of the control group but the extent of the sprouting was still quite extensive. The animals maintained at the highest level (28 μg CORT) were severely affected by the glucocorticoids. These animals showed a significant reduction in the lesion-induced growth. This reduction was observed as both a decrease in number of fibers and suppression of lateral progression of the growth response.

Again, constant elevated circulating CORT levels similar to those of normal aged animals significantly impaired the sprouting response in young animals. Accordingly, these same circulating CORT levels may have a potentially suppressive effect on axon sprouting in the aged brain. Such high levels of CORT may have effects on normal regeneration and synaptic turnover in the aged, thus contributing to the general decline in neuronal connectivity.

7. CONCLUSION

These experiments have described a number of important features about the CNS in the aged animal. There appears to be hormonal regulation of the synaptic remodeling which occurs following damage.

We have been unable to find any decrease in the density of synaptic contacts in the hippocampus as a function of age. This is not suprising since our experimentation demonstrates that the aged nervous system is capable of supporting reactive growth albeit at a slower rate. Occasionally we have observed a degenerating terminal in the neuropil of naive aged animals perhaps indicative of normal synaptic turnover as proposed by Cotman et al. (1981) and Hoff et al. (1981). We would expect the aged animal to be able to compensate for a very slow loss of synaptic contacts. The critical problem, as we see it, hinges on the ability of the aged brain to replace the lost terminals with ones that are functionally appropriate. The aged animal may not be capable of doing this and hence over time the circuitry of the hippocampus and other areas may become inappropriately re-wired. This change in the circuitry may have significant functional manifesta-tions and underlie some of the deficits observed in the behavior of the aged animal.

When the aged animals are placed in a lesion–recovery paradigm and com-pared with young adult animals, they may begin with a marked disadvantage. Since their anatomical connections are not identical, due in part to synapse loss and subsequent synaptic reacquisition as part of the normal aging process, the new neural connections formed as a result of the perturbation will be quite different. Depending upon the degree of natural synaptic turnover in the aged animal prior to the experimental lesion, the rewiring which occurs postopera-tively will be amplified. Subsequently, these neural connections will function very differently in the two age groups. Two additional factors should be taken into account. First, the aged animals progress in the replacement process at a much slower rate. This may account, in part, for their significant decrease or delay in recovery on some behavioral tasks. If a certain number of synaptic contacts are necessary to begin processing of a specific type of information, then a delay in obtaining that minimal critical number would delay the animal's ability to initially respond. While the aged animal may eventually be able to solve the task, it would do so only after a longer convalesence period. Perhaps, if aged animals were given this longer convalescence period before being retested, some types of behavioral deficits would be minimized. Second, a rewiring of the neuronal circuitry as a result of the aging process may force the aged animal to employ a different strategy than that employed by the younger animals (see Chapters 7 and 8). Consequently, since the older animals are required to solve problems when their rewiring is in a transient stage again, they may be forced to employ a different behavioral strategy. The lack of appropriate signals or the influence of too many irrelevant signals could force the animal to try a large number of different strategies. It is known, for instance, that young animals persist in a single strategy after trying out several. The aged animals appear easily distracted perhaps having trouble analyzing and attending to the most appropriate cues necessary to solve the task.

While the rate of the plasticity response following damage is altered in the aged population, there appears to be a physiological reason for this, and our recent work with hormones appears to be providing some clues. Glucocorticoids have a tremendous clinical reputation in the reduction and prevention of cerebral edema following brain damage. Even though we do not know the exact mecha-nism for this, it has been postulated that the glucocorticoids work through a membrane stabilization or by a suppressive response on fibroblast (Gray *et al.*, 1971).

Our current model for factors which affect reactive synaptogenesis in young and aged animals is shown in Fig. 12. In this model we have tried to capture the

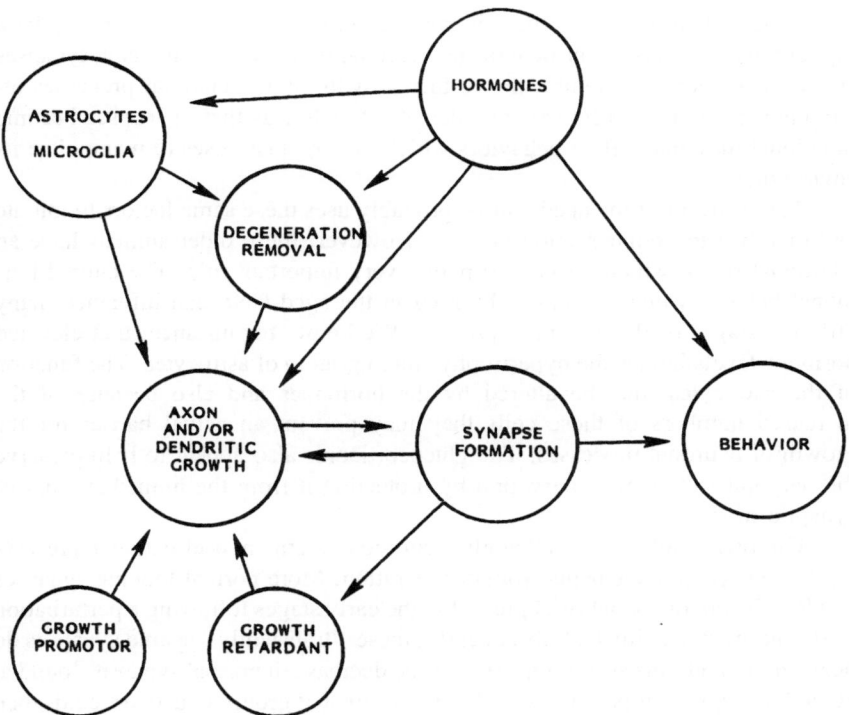

Figure 12. Model for the regulation of reactive synapse replacement in hippocampus. A loss of synapses, whether it be the result of a lesion or due to normal turnover, alters the balance of trophic factors which induces residual afferents to grow. A variety of factors influence the synaptic replacement. While degenerative debris may be a trigger mechanism, it is rapidly removed by the glial cells. The formation of new synapses also regulates the growth of neuronal processes which are under the control of hormonal influences. It may be these hormonal factors which slow the replacement process in the aged animals.

interaction of variables that experimentation has shown to be important in the reacquisition of lost synaptic contacts. Following synaptic loss, whether it be through a natural turnover of synapses or as the result of some specific type of perturbation, a number of factors are activated. A trophic substance is released which promotes growth from remaining cells in the denervated zone. This growth may be the result of a titration of growth retardants and growth promoting substances. Degeneration products appear prior to neuronal outgrowth and the removal of this debris involves both the astrocytes and microglia. Immediately following denervation the astrocytes appear to hypertrophy and engulf the degenerative debris, perhaps making space available for new synaptic connections. At the same time, the astrocytes may be communicating with the growing processes by directing them or aiding them in some fashion. The degeneration may be a trigger which signals the growth of the neuronal processes. As the new synapses are formed, they may inhibit the further growth of the neuronal processes attempting to attain a preset synaptic density. Finally, as these synapses become functional they may affect behaviors which the organism uses to manipulate its environment.

The neuropil of the aged animal probably uses these same factors to initiate and mobilize the reinnervation process. However, these older animals have an additional factor which appears to play a very important role. The altered hormonal balance which we have observed in the aged CNS can influence many different stages of the assembly process. We know, for instance, that elevated hormone levels induce the hypertrophy and migration of astrocytes. The function of the astrocytes may be altered by the hormones and also because of the increased numbers of these cells they may prevent an actual barrier for the growth of neuronal processes. The glucocorticoids also appear to help preserve the neuropil following denervation by protecting it from the immediate loss of synaptic input.

Our own results suggest that glucocorticoid treatment is effective in preserving the presynaptic elements from degeneration. More normal-looking synapses and less degenerative debris is present in the early stages following a perturbation in steroid-treated animals. Subsequently, these "transiently" spared synapses do degenerate, and thus account in part for the decrease in normal synapses found at 10 and 60 days postoperatively in the steroid-treated group. Glucocorticoids then appear to slow down the rate of degeneration and hence retard synapse replacement. Whether or not these steroids can permanently alter the magnitude of the response remains to be investigated. There does appear to be some precedent for this delay in degeneration. Hall et al. (1977) reported that glucocorticoids participated in the preservation of the degenerating motor neuron of the cat soleus muscle. Drakontides (1978) presented preliminary evidence suggesting that glucocorticoids may delay the onset of degenerative changes associated with denervation in rat motor neurons. Drakontides et al. (1981) also presented evidence that neuropathy caused by certain organophosphates like diisopropylfluorophosphate can be delayed with high doses of glucocorticoids.

Whatever the mechanisms are which regulate synaptogenesis in the aging animal, they do not totally suppress the reaction but allow for some compensation after the damage. This process probably plays a significant role in the ability of the aged animal to recover from damage and may be responsible, in part, for changes in behavior as a result of the normal aging process.

ACKNOWLEDGMENT.   This research was supported in part by a grant from the National Institute of Health NS16981.

REFERENCES

Bartus, C. A., and McNaughton, B. L., 1980, Spatial memory and hippocampal synaptic plasticity in senescent and middle-aged rats, in: *Psychobiology of Aging: Problems and Perspectives* (D. G. Stein, ed.), Elsevier/North-Holland, Amsterdam, pp. 253–272.

Bignami, A., and Ralson, J. H., 1969, The cellular reaction to Wallerian degeneration in the central nervous system of the cat, *Brain Res.* **13**:444–461.

Birren, J. W., and Shaie, K. W., 1977, *Handbook of Psychology of Aging,* Van Nostrand, New York.

Bondareff, W., and Geinisman, Y. 1976, Loss of synapses in the dentate gyrus of the senescent rat, *Am. J. Anat.* **145**:129–136.

Campbell, B. A., Krauter, E. E., and Wallace, J. W., 1980, Animal models of aging: Sensory-motor and cognitive function in aged rats, in: *Psychobiology of Aging: Problems and Perspectives* (D. G. Stein, ed.), Elsevier/North-Holland, Amsterdam, pp. 201–226.

Cotman, C. W., and Lynch, G. S., 1976, Reactive synaptogenesis in the adult nervous system: The effects of partial deafferentation on new synapse formation, in: *Neuronal Recognition* (S. Barondes, ed.), Plenum Press, New York, pp. 69–108.

Cotman, C. W., and Scheff, S. W., 1979, Synaptic growth in aged animals, in: *Aging, Physiology and Cell Biology of Aging* (A. Cherkin, C. W. Finch, N. Kharasch, T. Makinodan, F. L. Scott, and B. L. Strehler, eds.), Raven Press, New York, pp. 109–120.

Cotman, C. W., Nieto-Sampedro, M., and Harris, E. W., 1981, Synapse replacement in the adult nervous system of vertebrates, *Physiol. Rev.* **61**:684–784.

Dieringer, N., and Precht, W., 1977, Modification of synaptic input following unilateral labyrinthectomy, *Nature (London)* **269**:431–433.

Doty, B. A., 1966, Age differences in avoidance conditioning as a function of distribution of trials and task difficulty, *Gen. Psychol.* **109**:249–254.

Drakontides, A. B., 1978, Delay of denervation-degenerative changes in rat motor nerve terminals following glucocorticoid treatment, *Soc. Neurosci Abstr.* **4**:368.

Drakontides, A. B., Baker, T., and Riker, W. F., 1981, Glucocorticoid prevention of delayed organophosphorus neuropathy, *Anat. Rec.* **199**(3):73A.

Elias, P. K., and Elias, M. F., 1976, Effects of aging on learning ability: Contributions from the animal literature, *Exp. Aging Res.* **2**:165–186.

Field, P. M., Coldham, P. E., and Raisman, G., 1980, Synapse formation after injury in the adult rat brain: Preferential reinnervation of denervated fimbrial sites of axons of the contralateral fimbria, *Brain Res.* **189**:103–113.

Geinisman, Y., 1981, Loss of axon terminals contacting neuronal somata in the dentate gyrus of aged rats, *Brain Res.* **212**:136–139.

Geinisman, Y., and Bondareff, W., 1976, Decrease in the number of synapses in the senescent brain: A quantitative electron microscopic analysis of the dentate gyrus molecular layer in the rat, *Mech. Ageing Dev.* **5**:11–23.

Geinisman, Y., Bondareff, W., and Dodge, J. T., 1977, Partial deafferentation of neurons in the dentate gyrus of the senescent rat, *Brain Res.* **134**:541–545.

Gold, P. E., and McGaugh, J. L., 1975, Changes in learning and memory during aging, in: *Neurobiology of Aging* (J. M. Ordy and K. R. Brizzee, eds.), Plenum Press, New York, pp. 145–158.

Goldberger, M. E., and Murray, M., 1974, Restitution of function and collateral sprouting in the cat spinal cord: The deafferented animal, *J. Comp. Neurol.* **158**:37–54.

Goldowitz, D. and Cotman, C. W., 1980, Do neurotrophic interactions control synapse formation in the adult rat brain, *Brain Res.* **181**:325–344.

Goodrick, C. L., 1972, Learning by mature young and aged Wistar albino rats as a function of task complexity, *J. Gerontol.* **27**:353–357.

Gray, J. B., Pratt, W. B., and Aronow, L., 1971, Effects of glucocorticoids on hexose uptake by mouse fibroblasts in vitro, *Biochemistry* **10**:277–284.

Hall, E. D., Baker, T., and Riker, W. F., 1977, Glucocorticoid preservation of motor nerve function during early degeneration, *Ann. Neurol.* **1**:263–269.

Hasan, M., and Glees, P., 1973, Ultrastructural age changes in hippocampal neurons, synapses and neuroglia, *Exp. Gerontol.* **8**:75–83.

Hess, G. D., and Riegle, G. D., 1970, Adrenocortical responsiveness to stress and ACTH in aging rats, *J. Gerontol.* **25**:354–358.

Hoff, S. F., Scheff, S. W., Benardo, L. S., and Cotman, C. W., 1981, Lesion-induced synaptogenesis in the dentate gyrus of aged rats: I. Loss and reacquisition of normal synaptic density, *J. Comp. Neurol.* **205**:246–252.

Hoff, S. F., Scheff, S. W., and Cotman, C. W., 1982, Lesion-induced synaptogenesis in the dentate gyrus of aged rats: II. Demonstration of an impaired degeneration clearing response, *J. Comp. Neurol.* **205**:253–259.

Laatsch, R. H., and Cowan, W. M., 1967, Electron microscopic studies of the dentate gyrus of the rat: II. Degeneration of commissural afferents, *J. Comp. Neurol.* **130**:241–250.

Landfield, P. W., Rose, G., Sandles, L., Wohlstadter, T., and Lynch, G., 1977, Patterns of astroglial hypertrophy and neuronal degeneration in the hippocampus of aged memory-deficient rats, *J. Gerontol.* **32**:3–12.

Landfield, P. W., Waymire, J. C., and Lynch, G., 1978, Hippocampal aging and adrenocorticoids: Quantitative correlations, *Science* **202**:1098–1102.

Landfield, P. W., Wurtz, C., and Lindsey, J. D., 1979, Quantification of synaptic vesicles in hippocampus of aging rats and initial studies of possible relations to neurophysiology, *Brain Res. Bull.* **4**:757–763.

Loesche, J., and Steward, O., 1977, Behavioral correlates of denervation and reinnervation of the hippocampal formation of the rat: Recovery of alternation performance following unilateral entorhinal cortex lesions, *Brain Res. Bull.* **2**:31–39.

Loy, R., and Moore, R. Y., 1977, Anomalous innervation of the hippocampal formation by peripheral sympathetic axons following mechanical injury, *Exp. Neurol.* **57**:645–651.

Lynch, G., Gall, C., Rose, G., and Cotman, C. W., 1976, Changes in the distribution of the dentate gyrus associational system following unilateral or bilateral entorhinal lesion in the adult rat, *Brain Res.* **11**:57–71.

Lynch, G., Gall, C., and Cotman, C. W., 1977, Temporal parameters of axon "sprouting" in the brain of adult rats, *Exp. Neurol.* **54**:179–183.

Matthews, D. W., Cotman, C. W., and Lynch, G. S., 1976a, An electron microscopic study of lesion-induced synaptogenesis in the dentate gyrus of the adult rat: I. Magnitude and time course of degeneration, *Brain Res.* **115**:1–21.

Matthews, D. W., Cotman, C. W., and Lynch, G. S., 1976b, An electron microscopic study of lesion-induced synaptogenesis in the dentate gyrus of the adult rat: II. Reappearance of morphologically normal contacts, *Brain Res.* **115**:22–41.

Moore, R. Y., Bjorklund, A., and Stenevi, U., 1971, Plastic changes in the adrenergic innervation of the rat septal area in response to denervation, *Brain Res.* **33**:13–35.

Moore, R. Y., Bjorklund, A., and Stenevi, U., 1974, Growth and plasticity of adrenergic neurons, in: *The Neurosciences: Third Study Program* (F. O. Schmidt and F. B. Worden, eds.), MIT Press, Cambridge, pp. 961–977.

Murray, M., and Goldberger, M. E., 1974, Restitution of function of collateral sprouting in the cat spinal cord: The partially hemisected animal, *J. Comp. Neurol.* **158**:19–36.

Raisman, G., 1969a, A comparison of the mode of termination by hippocampal and hypothalamic afferents to the septal nuclei as revealed by electron microscopy of degeneration, *Exp. Brain Res.* **7**:317–343.

Raisman, G., 1969b, Neuronal plasticity in the septal nuclei of the adult rat, *Brain Res.* **14**:25–48.

Raisman ,G., and Field, P., 1973, A quantitative investigation of the development of collateral reinnervation after partial deafferentation of the septal nuclei, *Brain Res.* **50**:241–264.

Reiss, D. J., Ross, R. A., Gilad, G., and Joh, T. H., 1978, Reaction of central catecholaminergic neurons to injury: Model systems for studying the neurobiology of central regeneration and sprouting, in: *Neuronal Plasticity* (C. W. Cotman, ed.), Raven Press, New York, pp. 197–226.

Riegle, G. D., and Hess, G. D., 1972, Chronic and acute dexamethasone suppression of stress activation of the adrenal cortex in young and aged rats, *J. Neuroendocrinol.* **9**:175–187.

Rigter, H., Martinex, J. L., and Crabbe, J. C., 1980, Forgetting and other behavioral manifestations of aging, in: *Psychobiology of Aging: Problems and Perspectives* (D. G. Stein, ed.), Plenum Press, New York, pp. 161–176.

Rose, G., Lynch, G. S., and Cotman, C. W., 1976, Hypertrophy and redistribution of astrocytes in the deafferented dentate gyrus, *Brain Res. Bull.* **1**:87–92.

Scheff, S. W., and Cotman, C. W., 1977, Recovery of spontaneous alternation following lesions of the entorhinal cortex in adult rats: Possible correlation to axon sprouting, *Behav. Biol.* **21**:286–293.

Scheff, S. W., Benardo, L. S., and Cotman, C. W., 1977, Progressive brain damage accelerates axon sprouting in the adult rat, *Science* **197**:795–797.

Scheff, S. W., Benardo, L. S., and Cotman, C. W., 1978, Effect of serial lesions on sprouting in the dentate gyrus: Onset and decline of the catalytic effect, *Brain Res.* **150**:45–53.

Scheff, S. W., Benardo, L. S., and Cotman, C. W., 1980, Decline in reactive fiber growth in the dentate gyrus of aged rats compared to young adult rats following entorhinal cortex removal, *Brain Res.* **199**:21–38.

Scheff, S. W., Benardo, L. S., and Cotman, C. W., 1978, Decrease in adrenergic axon sprouting in the senescent rat, *Science* **202**:775–778.

Stenevi, U., and Bjorklund, A., 1978, A pitfall in brain lesion studies: Growth of vascular sympathetic axons into the hippocampus after fimbrial lesions, *Neurosci. Lett.* **7**:219–224.

Steward, O., Cotman, C. W., and Lynch, G. S., 1974, Growth of a new fiber projection in the brain of adult rats: Re-innervation of the dentate gyrus by the contralateral entorhinal cortex following ipsilateral entorhinal lesions, *Exp. Brain Res.* **20**:45–66.

Steward, O., Loesch, J., and Horten, W. C., 1977, Behavioral correlation of denervation and reinnervation of the hippocampal formation of the rat: Open field activity and cue utilization following bilateral entorhinal cortex lesions, *Brain Res. Bull.* **2**:41–48.

Steward, O., 1982, Assessing the functional significance of lesion-induced neuronal plasticity, *Int. Rev. Neurobiol.* **23**:197–253.

Tang, G., and Phillips, R., 1978, Some age-related changes in pituitary-adrenal function in the male laboratory rat, *J. Gerontol.* **33**:377–382.

West, J. R., Deadwyler, S. A., Cotman, C. W., and Lynch, G. S., 1975, Time-dependent changes in commissural field potentials in the dentate gyrus following lesions of the entorhinal cortex in adult rats, *Brain Res.* **97**:215–233.

Westrum, L. E., 1973, Early forms of terminal degeneration in the spinal trigeminal nucleus following rhizotomy, *J. Comp. Neurol.* **2**:189–215.

5

Relationship of the Raphe and Suprachiasmatic Nuclei to Serotonin Facilitation of Cyclic Reproductive Functions in Aging Female Rats

RICHARD F. WALKER

1. INTRODUCTION

The female reproductive system undergoes degenerative structural changes and it accumulates severe functional deficits during aging. Although other physiological systems also degenerate to varying degrees with advancing age, normal reproductive function ceases entirely and is replaced by some type of abnormal gonadal activity in the female (Aschheim, 1976). For example, in the aged rat two or three postreproductive syndromes are recognized including constant estrus (CE) and repetitive pseudopregnancy (PSP). Symptomatically, these conditions differ in terms of ovarian morphology and endocrinology. CE is characterized by polyfollicular ovaries lacking corpora lutea, while ovaries from rats showing PSP contain many corpora lutea and the number of growing and preovulatory follicles is greatly reduced. As the result of these structural changes, sex steroid production by the ovaries of CE, PSP, and normally cycling rats differ significantly. CE ovaries hyperproduce and secrete estrogen resulting in elevated serum levels of the steroid which persist chronically (Huang *et al.*, 1978). Since corpora lutea are lacking, progesterone production and serum con-

RICHARD F. WALKER • Department of Anatomy and Sanders–Brown Research Center on Aging, University of Kentucky Medical Center, Lexington, Kentucky 40536.

tent are profoundly depressed (Miller and Riegle, 1980). On the other hand, serum from PSP rats contains elevated levels of progesterone and relatively little estrogen. In contrast to these senile variations, ovaries from young rats showing regular estrous cycles produce sequentially estrogen and progesterone as correlates of stages of the cycle and of the periodic presence of graffian follicles and corpora lutea.

After many years of investigation, it is currently recognized that the primary lesion of aging in the female reproductive system leading to loss of cyclic ovarian function occurs within the brain and not in peripheral components of the hypothalamo-pituitary–ovarian axis. This is clearly demonstrated by the fact that reproductively nonfunctional ovaries and pituitaries taken from aged rats regain function when they are transplanted into young hosts (Peng and Huang, 1972). Furthermore, centrally acting drugs reinstate estrous cycles and ovulation in aged, previously acyclic animals suggesting that age changes in the brain lead to ovarian dysfunction. For example, monoamines and their precursors, such as L-dihydroxyphenylalanine (L-dopa), stimulate ovulation when administered systemically (Quadri et al., 1973). This effect is central since it persists when L-dopa is administered with drugs blocking its decarboxylation and transformation to catecholamines in the periphery (Linniola and Cooper, 1976). Furthermore, the central effect is localized to the rostral hypothalamus since placement of L-dopa into different regions of the brain reestablishes estrous cycles and ovulation only when the drug is implanted into the medial preoptic region (MPOA) (Cooper et al., 1979). This region of the hypothalamus apparently contains the neural circuitry responsible for generating periodic signals known to trigger the preovulatory surges of luteinizing hormone (LH), i.e., central events of the estrous cycle and a key factor in maintaining cyclic function in the female reproductive system. Thus, an anatomical locus for the primary age lesion may be within hypothalamic cells generating the neural stimulus for the preovulatory LH surge. As the neural signal degenerates and fails, the LH surge becomes unstable and finally ceases causing estrous cycles to arrest and CE or PSP to develop. In support of this hypothesis is the finding that electrolytic lesions of MPOA produce PSP (Clemens and Bennett, 1977), while electrolytic and pharmacologic lesions of the suprachiasmatic nuclear region (SNR) produce CE (Brown-Grant and Raisman, 1977; Walker et al., 1980). These treatments seem to produce their effects, at least in part, by affecting monoamine neurons since they are reversed by catecholamine (CA) or serotonin (5-HT) neuroleptics. For example, lergotrile mesolate, a dopamine receptor agonist reinstates cycles in rats made PSP by MPOA lesions (Clemens and Bennett, 1977). Furthermore, placement of p-chlorophenylalanine (pCPA) into SNR to block local 5-HT synthesis produces CE. This effect is reversed by 5-hydroxytryptophan (5-HTP), a precursor of the monoamine, suggesting that 5-HT in SNR is required for signaling the LH surge. This conclusion also derives from the fact that the facilitatory effect of 5-HT upon the LH surge requires the neural circuitry of the supra-chiasmatic nucleus (Coen and MacKinnon, 1980).

The present study investigates further how changes in serotoninergic neurons within the brain may be involved in age changes leading to loss of cyclic function in the female reproductive system. During earlier investigations, it was found that 5-HT metabolism in the hypothalamus becomes disturbed after approximately one year of age (Simpkins et al., 1977; Walker, 1980). These changes lead to perturbations of the normal circadian rhythm in hypothalamic 5-HT content (Walker, 1980) which may be an essential part of the neural mechanism regulating phasic LH secretion (Héry et al., 1976; Walker, 1980). It was proposed that part of the brain receiving 5-HT signals for LH release becomes hyposensitive to those stimuli and that the sensitivity change leads to an apparent diminution and ultimate loss of the signal. To test this hypothesis, old rats showing CE were made supersensitive to serotoninergic signals by pharmacological methods (Walker, 1983a). It is known that chronic interruption of neurotransmission within monoamine circuits make systems dependent upon them to show potentiated responses when the signals are restored. Thus, old CE rats were made hypersensitive to 5-HT signals by treatment with pCPA. The drug blocks 5-HT synthesis, thereby depressing neuronal content of the neurotransmitter and blocking transmission. After two days, the pharmacologic block was circumvented with 5-HTP, thus restoring 5-HT synthesis. Luteinizing hormone surges were reinstated by the pCPA + 5-HTP treatment in CE rats suggesting that neurochemical changes involving 5-HT may be in part responsible for loss of LH surges with advancing age. To determine if these changes were occurring within discrete regions of the central serotoninergic system, 5-HT metabolism was altered within the SCN and raphe nuclei with drugs which were applied locally. The aim was to determine if and how changes in serotonin metabolism in suprachiasmatic nucleus (SCN) and/or raphe lead to cycle irregularity and degeneration of female reproductive function during aging.

2. TEMPORAL PATTERNS OF HYPOTHALAMIC 5-HT AND CYCLIC REPRODUCTIVE FUNCTION

Serotonin facilitates the LH surge (Héry et al., 1976; Coen and MacKinnon, 1979) by some yet unknown mechanism. However, recent findings suggest that phasic secretion of LH is functionally related to the circadian rhythm in hypothalamic 5-HT content (Walker, 1980, 1983b) first described by Quay (1968). It appears that the postacrophase component of the 5-HT rhythm reflects the intensity of a 5-HT signal contributing to the magnitude of the LH surge. Thus, it was shown in young animals that large differences between peak and nadir values for 5-HT content correlate with large LH surges, whereas smaller 5-HT differences were related to attenuated LH surges. Furthermore, LH surges did not occur, indicating that sensitivity to the positive feedback effects of estrogen were lost, when the 5-HT circadian rhythm was abolished experimentally (Walker, 1983b). The significance of these findings to age changes in reproductive function lies in the fact that aged female rats in CE lack the 5-HT circadian rhythm (Walker, 1980). This aberration in the temporal pattern of 5-HT metabolism seems also to be related to loss of LH surges in old rats, since a combination of drugs inducing supersensitivity in serotoninergic systems and reestablishing the normal pattern of hypothalamic 5-HT metabolism reinstated LH surges in previously acyclic CE rats (Walker, 1983a). Thus, age changes in the temporal pattern of hypothalamic 5-HT metabolism seem to be related to the facilitatory effect of 5-HT on the LH surge.

The SCN is the major controller of circadian organization in the rat (Rusak and Zucker, 1979). Furthermore, this hypothalamic structure is rich in 5-HT axon terminals (Saavedra et al., 1974) whose metabolic activity is correlated with estrogen-induced LH surges (Héry et al., 1982). Since the phasic secretion of LH is a circadian rhythm (Chazal et al., 1977) which is lost when the SCN is destroyed (Brown-Grant and Raisman, 1977), the relationship between 5-HT as a pacemaker for LH surges emerges. In fact, it has been previously shown that the neural circuits involved in the facilitatory effect of 5-HT on LH secretion resides in SCN (Coen and MacKinnon, 1980). Therefore, 5-HT content in SCN was measured to determine if, as in the hypothalamus, a circadian pattern exists in this nucleus proper, and also, if the pattern differs between cyclic and CE rats.

Young rats (4–5 months old) were killed at different times on the day of estrus. Middle-aged rats (12–14 months old) in CE were killed at comparable times. Some rats from both groups received single intraperitoneal injections of pargyline (75 mg/kg) 30 min before death. The drug was used to determine differential synthesis of 5-HT throughout the day. Pargyline blocks monoamine oxidase; therefore, a greater accumulation of 5-HT after pargyline at one time point versus another would suggest greater synthesis at the first time point. After decapitation, the brains were rapidly removed from the skull and stored at $-80°C$ until the SCN were dissected according to the method of Palkovits (1973). Pooled nuclear punches were sonicated in 0.1 N perchloric acid and centrifuged, and the supernatants containing extracted serotonin were assayed radioenzymatically (Saavedra et al., 1973). A fraction of the homogenate was used for protein determination. Table 1 shows that young rats killed on proestrus had similar daily patterns of SCN:5-HT as in the hypothalamus proper, i.e., highest levels and greatest synthesis occurred at approximately midpoint in the photophase, while baseline levels were reached several hours after dark. On the other hand, the 5-HT:SCN circadian rhythm was absent in CE rats, and the temporal profile again paralleled that of the hypothalamus from CE rats (Walker, 1980). In this case, periodic changes in 5-HT were not associated with phases of the photoperiod. Instead, 5-HT levels were uniform throughout the day, as the result of

Table 1. Daily Patterns of 5-HT Content in Hypothalamus and SCN from Rats with Regular Estrous Cycles or Constant Estrus[a]

	Serotonin content			
	Estrous cycles		Constant estrus	
Time	Hypothalamus (pg/mg)	SCN (ng/mg prot)	Hypothalamus (pg/mg)	SCN (ng/mg prot)
0900	919 ± 53	8.53 ± 0.79	1096 ± 114	11.35 ± 1.49
1400	1291 ± 122[b]	16.21 ± 2.57[c]	1215 ± 153	13.56 ± 1.70
1900	1055 ± 101	9.61 ± 1.04	1100 ± 69	10.74 ± 0.93
2400	814 ± 71	10.20 ± 1.15	1144 ± 111	11.00 ± 1.49

[a] Aged rats in CE or young rats with regular 4-day estrous cycles were killed at various times throughout the day for determination of hypothalamus or SCN 5-HT content. Daily fluctuations in SCN and hypothalamus were similar in SCN and hypothalamus. Furthermore, the daily changes were lost at both sites when rats entered constant estrus. Values represent means ± SEM for determination on six rats per group.
[b] $P < 0.05$.
[c] $P < 0.01$ when compared with 0900 hr.

increased production during the scotophase. Thus, loss of the SCN:5-HT circadian rhythm in CE rats correlates with loss of the hypothalamic 5-HT rhythm in the same animals. Since the hypothalamic 5-HT rhythm is a component of the facilitatory effect of 5-HT upon LH surges (Walker, 1980, 1983b), then perhaps neurochemical changes involving 5-HT in SCN, the circadian pacemaker of the rat brain, are associated with loss of cycles in aging rats.

To further evaluate this hypothesis, young rats were made CE by electrolytic destruction of SCN. The purpose was to determine if SCN circuits controlling estrous cyclicity also regulate the hypothalamic 5-HT rhythm. Since 5-HT afferents innervating the hypothalamus emerge from cells in the raphe nuclei (Moore *et al.*, 1978), lesions were also made in this area to determine the effect on estrous cycles and hypothalamic circadian rhythms.

Lesions of the SCN or dorsal raphe (RN) were placed stereotaxically in rats anesthetized with chloral hydrate. Following a 30-day recovery period from surgery, daily vaginal smears were taken to determine if the lesions were properly placed, especially those in SCN. Rats with presumptive SCN lesions were monitored until they showed persistent vaginal cornification, indicating that the SCN were destroyed (Brown-Grant and Raisman, 1977). No estrous cycle marker exists for destruction of the RN, therefore, all rats receiving midbrain lesions were used, and placement of raphe lesions were confirmed histologically at necropsy.

Sixty days after surgery, rats receiving SCN or RN lesions were differentiable by their vaginal smear patterns, as well as by their pattern of hypothalamic serotonin content throughout a 24-hr period. Rats with SCN lesions entered constant vaginal estrus without return to regular cycles for up to 60 days after surgery. On the other hand, rats with RN lesions showed periods of prolonged diestrus followed by cycle irregularity and finally a return to regular cycling after 35–50 days postsurgery. As shown in Fig. 1, hypothalamic 5-HT content at midpoint in the photophase in rats with discrete SCN lesions was slightly lower than sham-operated controls, though the differences were not significant. However, hypothalamic 5-HT content of SCN-lesioned rats was significantly higher ($P < 0.05$) during the dark phase of the photoperiod. On the other hand, RN lesions reduced hypothalamic 5-HT content throughout the period of examination. However, despite the overall reduction, a diurnal rhythm in hypothalamic 5-HT content persisted, having the same pattern, though attenuated, as controls. Rats with RN lesions showed regular estrous cycles; however, the magnitude of their preovulatory LH surges were attenuated as a correlate of the depressed hypothalamic 5-HT rhythm (Fig. 2).

These data show that destructive lesions of the SCN in young female rats produce changes in reproductive function and hypothalamic 5-HT metabolism resembling those occurring spontaneously in their aging counterparts. On the

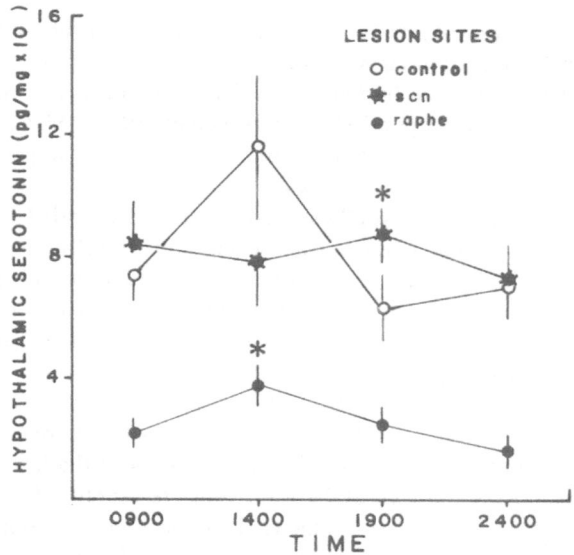

Figure 1. Effect of lesions of the suprachiasmatic (SCN) or raphe nucleus on hypothalamic serotonin (5-HT) content. Electrolytic destruction of SCN (✸) abolished daily fluctuations in hypothalamic 5-HT content. *, $P < 0.05$ compared with control at 1900 hr. Lesions of the raphe (●) significantly ($P < 0.01$) lowered hypothalamic 5-HT content at all time points compared with controls and SCN-lesioned animals. However, hypothalamic 5-HT content was significantly (*, $P < 0.05$) higher at 1400 compared with 2400 hr in rats with raphe lesions.

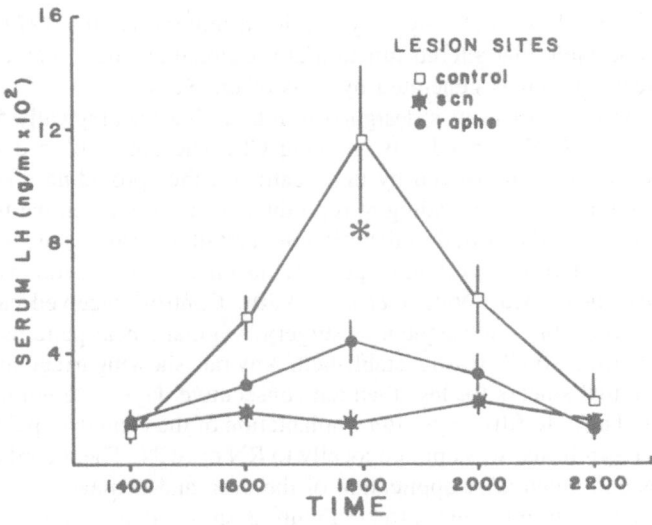

Figure 2. Effect of lesions in suprachiasmatic (SCN) or raphe (RN) nuclei on estrogen-induced LH surges in ovariectomized rats. LH surges were absent in rats with SCN (✹) lesions. On the other hand LH surges occurred in rats with RN (●) lesions, although peak levels of serum LH (1800 hr) were significantly (*, $P < 0.01$) higher in controls (□).

other hand, rats with RN lesions continued to show regular estrous cycles, although the amount of LH released on proestrus was reduced. This finding agrees with a previous report showing that the magnitude of the hypothalamic 5-HT rhythm is directly correlated with the size of the LH surge (Walker, 1983b). Furthermore, these findings suggest that the RN either provides a source of 5-HT for the hypothalamus or directs its synthesis therein, but that it does not dictate the circadian pattern of 5-HT metabolism within the hypothalamic tissue. This function seems to reside with the SCN, which acts as a circadian pacemaker for the 5-HT rhythm as well as many other functions previously described (see Rusak and Zucker, 1979). Thus, these experiments indicate that loss of circadian organization possibly involving the SCN and hypothalamic 5-HT metabolism may be involved in reproductive dysfunction with advancing age in female rats. However, it is not clear whether the changes leading to temporal dysfunction are primary to the SCN or secondary to changes at other loci such as the RN. In earlier studies (Labhsetwar, 1972) it was shown that reproductive dysfunction and loss of hypothalamic 5-HT rhythms could be induced in young rats by administering 5-HTP, a precursor to 5-HT. Presumably, 5-HT hypersynthesis resulting from this treatment "swamped out" the circadian pattern of hypothalamic 5-HT metabolism and also blocked LH surges resulting in loss of ovulation.

Since aged rats show 5-HT hypersynthesis (Simpkins *et al.*, 1977; Walker, 1980), it is possible that altered function at noncircadian sites, such as the RN, distorts circadian signals generated by cells of the SCN.

The next experiment was designed to retard pharmacologically 5-HT synthesis at RN or SCN in aged rats showing CE. The aim was to determine if estrous cycles can be reinstated by this treatment, thus providing a method for analysis of the mechanism leading to reproductive senescence in the female rat. Double-barrel cannulae for administration of crystalline drugs to the RN or SCN were implanted stereotaxically into aged CE rats under chloral hydrate anesthesia as previously described (Walker *et al.*, 1980). Controls received cannulae in frontal cortex or pons. Subsequent to surgery, vaginal smear patterns were followed to determine if CE was reestablished. Any rats showing irregular cycles or cornified vaginal smears for less than ten consecutive days were not used in the experiment. Thirty to fifty days after implantation of the cannulae, pCPA, which blocks 5-HT synthesis, was applied locally to RN or SCN. The record of vaginal smears was continued after application of the drug and compared with the pre-drug history to determine any effect. Figure 3 shows that estrous cycles were reinstated in rats receiving RN, pCPA implants. Within one day, vaginal smears were of the diestrous type. A period of irregular cycles followed for approximately two weeks, after which regular 4-day cycles were re-established. Application of leucine to RN as a control for pCPA had no effect on vaginal smears, indicating that inhibition of 5-HT at this locus altered the pattern of constant vaginal cornification. pCPA at cortical and pontine sites also had no effect on reproductive cycles, indicating that the effect of pCPA was site-specific. pCPA applied to SCN also altered vaginal smears, but regular cycles were not re-established by drug treatment at this site. A period of prolonged diestrus usually followed pCPA application to SCN, then the rats returned to CE. No rats with pCPA/SCN implants showed regular cycles. As seen in Table 2, placement of pCPA in RN or SCN affected 5-HT levels and metabolic patterns similar to electrolytic lesions at these sites. These findings suggest that regulatory dysfunction(s) occurring at the level of the RN is responsible for perturbations in hypothalamic 5-HT metabolism and associated disorders of cyclic reproductive function typically occurring in aging female rats. It appears that the midbrain changes depress the ability of SCN to provide circadian organization, at least with regard to patterns of hypothalamic 5-HT metabolism and estrous cycles. This is evident in the fact that the 5-HT circadian rhythm and estrous cycles are restored after depression of 5-HT synthesis by pCPA in RN.

Finally, it was of interest to determine whether rostral hypothalamic 5-HT neurons purportedly involved in facilitation of the LH surge (Walker *et al.*, 1980; Coen and MacKinnon, 1980) are still capable of subserving this function in the CE rats which have become insensitive to the positive-feedback effects of estrogen. Previously, it was shown that pharmacologic induction of supersen-

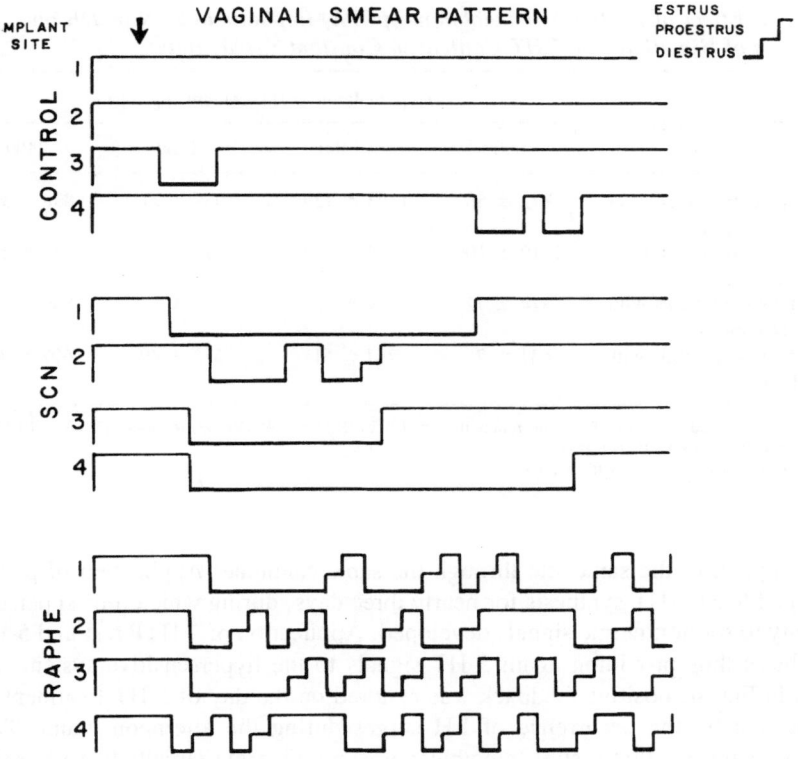

Figure 3. Restored vaginal cycles in constant estrus rats after placement of *p*-chlorophenylalanine (pCPA) in the dorsal raphe nucleus. ↓ , Time of pCPA application.

sitivity in serotoninergic neurons by systemic administration of pCPA + 5-HTP restored positive feedback in aged rats showing CE (Walker, 1983a). However, the effect of these drugs was not localized in that study. Therefore, CE rats receiving hypothalamic drug cannulae implanted as above into the medial preoptic/suprachiasmatic region. The animals were also ovariectomized and immediately received subcutaneous implants of silastic tubing containing estradiol 17β (Legan *et al.*, 1975). The exogenous steroid was provided to sustain a constant source of estrogen, which would have been interrupted if provided by the ovaries by the trauma of surgery. Three days after surgery, blood was withdrawn from the tail vein during the afternoon hours, when LH surges normally occur in estrogen-implanted rats, to determine serum LH content as background data prior to drug treatment. On day 4, pCPA was applied through the internal drug-delivery cannulae to MPOA/SCN. Sixty-seven hours later, at 1100 hr, 5-HTP

Table 2. Effect of p-Chlorophenylalanine (pCPA) Implants in SCN or DR on Patterns of Hypothalamic 5-HT Content in Constant Estrus Rats[a]

Treatment site and category	Hypothalamic 5-HT content (pg/mg)			
	0900	1400	1900	2400
Young control (regular estrous cycles)	872 ± 56	1193 ± 139[b]	959 ± 71	898 ± 97
Aged control (constant estrus)	1019 ± 100	1115 ± 122	1179 ± 96	1050 ± 120
Aged constant estrus with SCN implant	814 ± 72	879 ± 93	944 ± 57	899 ± 100
Aged constant estrus with DR implant	631 ± 7	787 ± 51[b]	673 ± 49	697 ± 70

[a]Controls received implants of leucine rather than pCPA in SCN or DR. Values represent means ± SEM for determinations in six rats per group.
[b]$P < 0.05$ when compared with 0900.

was applied to the same site through the same cannulae. Application of pCPA blocked local 5-HT synthesis for nearly three days, during which time supersensitivity to serotoninergic signals developed. Application of 5-HTP restored 5-HT synthesis thus providing again 5-HT signals to the hypersensitive system. As seen in Fig. 4, positive feedback was restored on the day of 5-HT treatment as evidenced by the occurrence of LH surges during the afternoon hours. This finding suggests that rostral hypothalamic serotoninergic circuits that normally facilitate positive feedback events such as the LH surge become nonfunctional with advancing age, but still exist and can be reactivated by drugs which make them supersensitive to serotoninergic signals.

In conclusion, reproductive dysfunction in aging female rats may be related in part to metabolic changes in serotonin occurring in midbrain sites and affecting the activity of diencephalic structures which normally play a role in maintaining estrous cycles and directing LH positive-feedback responses to estrogen. It is clear that the temporal organization and integration upon which female reproductive function depends is lost during aging. However, the primary lesion causing loss of temporal order may not reside within the SCN, at least with respect to 5-HT and the facilitatory contribution it makes to the phasic secretion of LH. It has been shown that the SCN is essential for maintaining hypothalamic 5-HT rhythms and LH surges, but that the function of this nucleus can be modulated by input from the RN. Thus, circadian rhythms and estrogen feedback responses can be abolished by excessive serotoninergic input from RN to the hypothalamus. This is shown by the fact that inactive rostral hypothalamic circuits can be reactivated by drug-induced 5-HT supersensitivity, and cyclic function can be

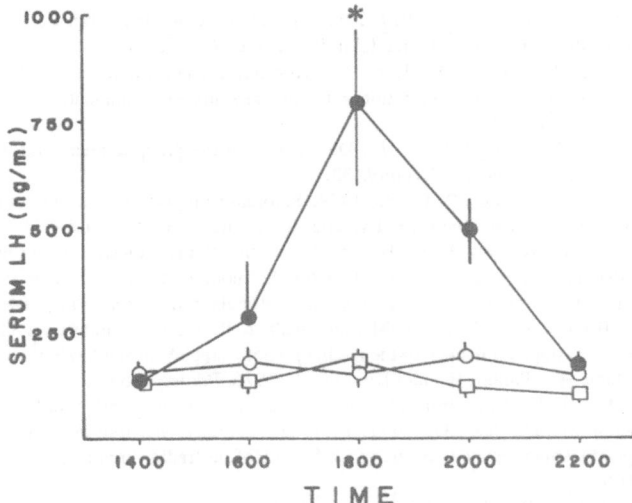

Figure 4. Restored positive-feedback responses in estrogen-treated constant estrus rats after *p*-chlorophenylalanine (pCPA) and 5-hydroxytryptophan (5-HTP) were applied to the suprachiasmatic nuclear area. Supersensitivity to serotonin signals was induced by pretreatment with pCPA. Sixty-seven hours later, rats received 5-HTP. Controls received leucine instead of pCPA or 5-HTP. □, pCPA + leucine; ○, leucine + 5HTP; ●, pCPA + 5HTP; *, $P < 0.001$ when compared with either control at same time point.

restored by retarding 5-HT synthesis in RN. Though it still is unknown whether 5-HT synthesized in RN is transported to hypothalamic sites in excess, or if genetic changes in 5-HT perikarya in RN are expressed as hypersynthesis of 5-HT at their hypothalamic axon terminals, it seems clear that these changes are associated with advancing age and have disruptive effects upon female reproductive function. Thus, changes in the properties of RN and 5-HT neurons may represent an important part of the mechanism for reproductive system aging in certain female animals.

ACKNOWLEDGMENTS. This study was supported in part by research grant No. AG 02867 from the National Institute on Aging. Thanks are expressed to Ms. Audrey Neilsen for typing and final preparation of the manuscript.

REFERENCES

Aschheim, P., 1976, Aging in the hypothalamic-hypophyseal-ovarian axis in the rat, in: *Hypothalamus, Pituitary and Aging* (A. V. Everitt and J. A. Burgess, eds.), Charles C. Thomas, Springfield, Illinois, pp. 376–418.

Brown-Grant, K., and Raisman, G., 1977, Abnormalities in reproductive function associated with destruction of the suprachiasmatic nuclei in female rats, *Proc. R. Soc. (London)* **198**:279–296.

Chazal, G., Faudon, M., Gogan, F., Héry, M., Kordon, C., and LaPlante, E., 1977, Circadian rhythm of luteinizing hormone secretion in the ovariectomized rat implanted with oestradiol, *J. Endocrinol.* **75**:251–260.

Clemens, J. A., and Bennett, D. R., 1977, Do changes in the preoptic area contribute to loss of cyclic endocrine function? *J. Gerontol.* **32**:19–24.

Coen, C. W., and MacKinnon, P. C. B., 1979, Serotonin involvement in the control of phasic luteinizing hormone release in the rat: Evidence for a critical period, *J. Endocrinol.* **82**:105–113.

Coen, C. W., and MacKinnon, P. C. B., 1980, Lesions of the suprachiasmatic nuclei and the serotonin-dependent phasic release of luteinizing hormone in the rat: Effects of drinking rhythmicity and on the consequences of preoptic area stimulation, *J. Endocrinol.* **84**:231–236.

Cooper, R. L., Brandt, S. J., Linniola, M., and Walker, R. F., 1979, Induced ovulation in aged female rats by L-dopa implants into the medial preoptic area, *Neuroendocrinology* **28**:234–240.

Héry, M., LaPlante, E., Pattou, E., and Kordon, C., 1976, Participation of serotonin in the phasic release of LH. I. Evidence from pharmacological experiments, *Endocrinology* **99**:496–503.

Héry, M., Faudon, M., Dusticier, G., and Héry, F., 1982, Daily variations in serotonin metabolism in the suprachiasmatic nucleus of the rat: Influence of oestradiol impregnation, *J. Endocrinol.* **94**:157–166.

Huang, H. H., Steger, R. W., Bruni, J. F., and Meites, J., 1978, Patterns of sex steroid and gonadotropin secretion in aging female rats, *Endocrinology* **103**:1855–1861.

Labhsetwar, A. P., 1972, Role of monoamines in ovulation: Evidence for a serotoninergic pathway for inhibition of spontaneous ovulation, *J. Endocrinol.* **54**:269–274.

Legan, S. J., Coon, G. A., and Karsch, F. J., 1975, Role of estrogen as initiator of daily LH surges in the ovariectomized rat, *Endocrinology* **96**:50–56.

Linniola, M., and Cooper, R. L., 1976, Reinstatement of vaginal cycles in aged female rats, *J. Pharm. Exp. Ther.* **199**:477–482.

Miller, A. E., and Riegle, G. D., 1980, Temporal changes in serum progesterone in aging female rats, *Endocrinology* **106**:1579–1583.

Moore, R. Y., Halaris, A. E., and Jones, B. E., 1978, Serotonin neurons of the midbrain raphe: Ascending projections, *J. Comp. Neurol.* **180**:417–432.

Palkovits, M., 1973, Isolated removal of hypothalamic and other brain nuclei of the rat, *Brain Res.* **59**:449–450.

Peng, M. T., and Huang, H. O., 1972, Aging of hypothalamic-pituitary-ovarian function in the rat, *Fertil. Steril.* **23**:535–542.

Quadri, S. K., Kledzik, G. S., and Meites, J., 1973, Reinitiation of estrous cycles in old constant-estrous rats by central-acting drugs, *Neuroendocrinology* **11**:248–255.

Quay, W. B., 1968, Differences in circadian rhythms in 5-hydroxytryptamine according to brain region, *Am. J. Physiol.* **215**:1448–1453.

Rusak, B., and Zucker, I., 1979, Neural regulation of circadian rhythms, *Physiol. Rev.* **59**:449–526.

Saavedra, J. M., Brownstein, M., and Axelrod, J., 1973, A specific and sensitive enzymatic-isotopic microassay for serotonin in tissues, *J. Pharm. Exp. Ther.* **186**:508–515.

Saavedra, J. M., Palkovits, M., Brownstein, M. J. and Axelrod, J., 1974, Serotonin distribution in the nuclei of the rat hypothalamus and preoptic area, *Brain Res.* **77**:157–163.

Simpkins, J. W., Mueller, G. P., Huang, H. H., and Meites, J., 1977, Evidence for depressed catecholamine and enhanced serotonin metabolism in aging male rats: Possible relation to gonadotropin secretion, *Endocrinology* **100**:1672–1678.

Walker, R. F., 1980, Serotonin circadian rhythm as a pacemaker for reproductive cycles in female rats, in: *Progress in Psychoneuroendocrinology* (F. Brambilla, G. Racagi, and D. deWied, eds.), Elsevier/North-Holland, Amsterdam, pp. 591–600.

Walker, R. F., 1983a, Reinstatement of LH surges by serotonin neuroleptics in aging, constant estrous rats, *Neurobiol. Aging* **3**:253–257.

Walker, R. F., 1983b, Quantitative and temporal aspects of serotonin's facilitatory action on phasic secretion of LH in female rats, *Neuroendocrinology* **36**:468–474.

Walker, R. F., Cooper, R. L., and Timiras, P. S., 1980, Constant estrus: Role of rostral hypothalamic monoamines in development of reproductive dysfunction in aging rats, *Endocrinology* **107**:249–255.

6

Behavioral Consequences of Neuronal Plasticity following Injury to Nigrostriatal Dopaminergic Neurons

JOHN F. MARSHALL

1. INTRODUCTION

Damage to the central nervous system (CNS) can lead to behavioral abnormalities that often abate as the time since injury increases. This phenomenon, recovery of function, has been observed since the relationship between the nervous system and behavior was first recognized (see Rosner, 1974). During the past decade, interest in this recovery process has been renewed by advances in our knowledge of neuronal plasticity in the mammalian CNS. Far from being the static organs once envisaged, the brain and spinal cord can undergo fundamental cellular changes after injury.

Several types of neuronal adaptations to brain damage have been identified. First, intact axons can sprout collateral branches that provide a reinnervation of target sites vacated by degenerating axon terminals (Liu and Chambers, 1958; reviewed in Cotman, 1978). Also, severed axons can regrow processes from their proximal stumps (Ramon y Cajal, 1928; reviewed in Guth, 1974). In the mammalian CNS these growth processes rarely reach the site of original innervation; however, in some special circumstances they may (Nygren *et al.*, 1971, 1974). Third, surviving neurons within injured neural systems can alter their rate

JOHN F. MARSHALL • Department of Psychobiology, University of California, Irvine, Irvine, California 92717.

of cellular metabolism. For example, their rate of neurotransmitter synthesis and release can increase (Hefti *et al.*, 1980; Acheson *et al.*, 1980; Agid *et al.*, 1973; Sharman *et al.*, 1967) as can the apparent density of receptor sites for the transmitter (Creese and Snyder, 1979; Creese *et al.*, 1977; Sporn *et al.*, 1976; Nelson *et al.*, 1978; Westlind *et al.*, 1981). Fourth, nervous system injury may unmask synapses that have been latent (i.e., physiologically silent), perhaps through disinhibition (Millar *et al.*, 1976).

2. APPROACHES TO STUDYING THE NEURAL EVENTS MEDIATING BEHAVIORAL RECOVERY

The challenge to neuroscientists concerned with recovery of function after CNS injury is to demonstrate a causal relationship between one or more types of neuronal plasticity and behavioral restoration. This can be accomplished only by studying specific model systems of brain injury where both the behavioral and neurobiological consequences of the damage can be determined concurrently. The usefulness of any model system depends upon the satisfaction of three criteria:

1. Damage to the CNS region under investigation results in a well defined behavioral impairment followed by postoperative improvement. The time course and other behavioral characteristics of the recovery sequence are well characterized.
2. The behavioral abnormality is associated with damage to one (or more) identified class of neurons. Ideally, the extent of injury to this identified population of cells should be quantitatively related to the severity and duration of the ensuing behavioral impairments.
3. The types of neuronal plasticity that occur after the injury should either be known or be amenable to study using available techniques.

If these criteria are met, then the relationship of one or more types of neuronal plasticity to the recovery process can be tested by determining the correspondence between the postoperative time courses of the behavioral and cellular events. Further tests necessitate identifying experimental interventions that alter the time course or extent of the cellular changes in order to determine any effects on the behavioral recovery.

Several model systems of recovery from CNS injury currently under investigation appear to offer particular promise for identifying the neural events underlying behavioral restoration. These include the recovery of reflex function after spinal cord injury (Murray and Goldberger, 1974), the improvement in alternation performance that follows unilateral entorhinal cortex damage (Loesche and Steward, 1977), the return of spatial maze learning after damage to the septo-hippocampal projection (Low *et al.*, 1981), and the recovery of ingestive behavior (Stricker and Zigmond, 1976) and sensorimotor skills (Marshall, 1980) subsequent to damaging central dopaminergic projections.

During the past several years my associates and I have investigated the impairment in somatosensory localization and its recovery that occurs after damage to the nigrostriatal dopamine-containing projection of the rat. The goal of this chapter is to review our progress on this problem in order to illustrate the general approaches available for studying the neural events mediating recovery.

3. INATTENTION TO STIMULI

In 1971, Blair Turner, Philip Teitelbaum, and I found that an electrolytic lesion centered in the lateral hypothalamic area of the rat results in a syndrome of neglect, or inattention to stimuli. Although intact rats localize somatosensory, olfactory, or moving visual stimuli by turning their heads toward these stimuli, rats with lateral hypothalamic damage do not. Whereas bilateral injury produces bilateral symptoms, the impairment in localization following unilateral damage is restricted to stimuli impinging on the contralateral body side. Animals with such damage typically recover from these acute sensorimotor impairments (Marshall *et al.*, 1971).

The inattention to stimuli seen after lateral hypothalamic lesions appears not to be due to injuring hypothalamic neurons, but instead results from the interruption of passing axons. After Ungerstedt (1971a) described the ascending dopaminergic projection passing through the lateral hypothalamus en route to forebrain terminal sites, we found that an inattention to somatosensory (Fig. 1), visual, and olfactory stimuli occurs after 6-hydroxydopamine (6-OHDA) injections along the course of these ascending axons (Marshall *et al.*, 1974). As is the case following lateral hypothalamic lesions, these impairments are restricted to the contralateral body side after unilateral injury. When recovery occurs, as it frequently does, its temporal and somatotopic organization is highly stereotyped. The recovery of somatosensory localization, for example, begins when touch of the snout elicits turning of the head in the direction of the stimulus. The region of the body surface where touch evokes orientation subsequently spreads to include more caudal and dorsal regions of the affected body surface (Fig. 2). This recovery begins on day 4 or 5 postoperatively and is maximal by day 21 (Fig. 3).

4. NEGLECT AFTER NIGROSTRIATAL INJURY AS A MODEL SYSTEM

Our research has focused on the recovery of contralateral somatosensory localization that occurs after unilateral damage to the ascending dopaminergic projection in the rat. The orientation to tactile stimuli has proven to be especially

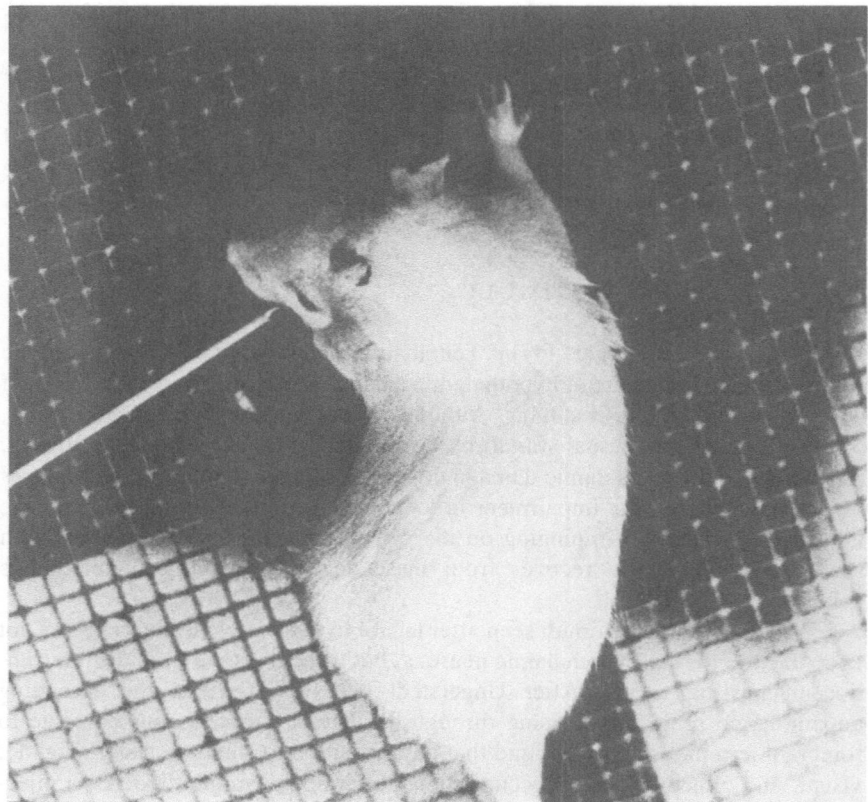

Figure 1. Contralateral somatosensory inattention after injection of 6-OHDA into the ventral tegmentum of the left hemisphere.

useful because the extent of localization at each body region and the somatotopic spread of sensitivity can be quantified (Marshall and Teitelbaum, 1974). Rats with unilateral damage have been used because the two body surfaces or two brain hemispheres of each animal can be compared with each other. However, before using this system to study the synaptic events underlying recovery of function, it is necessary to determine whether the impairments in somatosensory localization are due specifically to damaging the dopaminergic innervation of the neostriatum. In particular, two issues need to be addressed.

First, lateral hypothalamic electrolytic lesions or injections of 6-OHDA along the ascending dopaminergic axons damage dopamine-containing terminals throughout the forebrain. These axons innervate not only the neostriatum, but forebrain limbic structures [e.g., nucleus accumbens septi (NAS), olfactory tu-

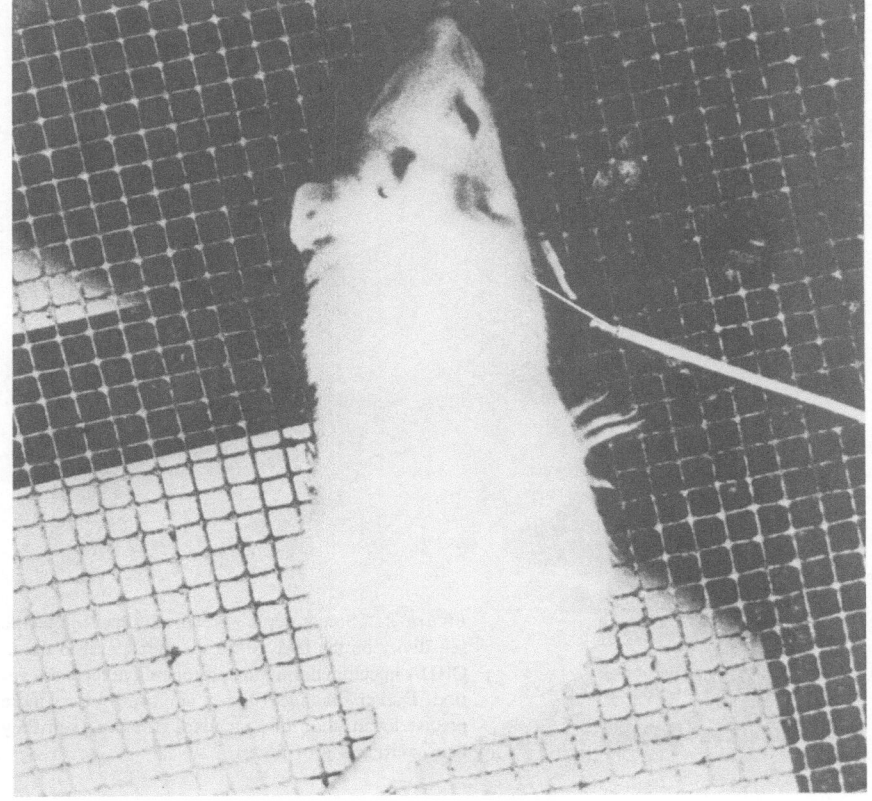

Figure 1 (continued)

bercle (OT), and lateral septum] and cortical regions as well (Ungerstedt, 1971a; Fallon and Moore, 1978; Lindvall and Björklund, 1974, 1978). What is the separate contribution of the neostriatal dopaminergic terminals to this behavior?

Second, although 6-OHDA is a relatively selective catecholamine neurotoxin, after its intracerebral injection at least a small region of nonspecific injury occurs at the injection tip (Hökfelt and Ungerstedt, 1973). Do 6-OHDA injections induce inattention to stimuli not because they interrupt brain dopaminergic projections but because they injure other systems nearby? As described below, the results of several experiments indicate that the impairment in somatosensory localization is attributable specifically to damaging the dopamine-containing afferents to the neostriatum.

After 6-OHDA injections that result in a substantial loss of dopaminergic

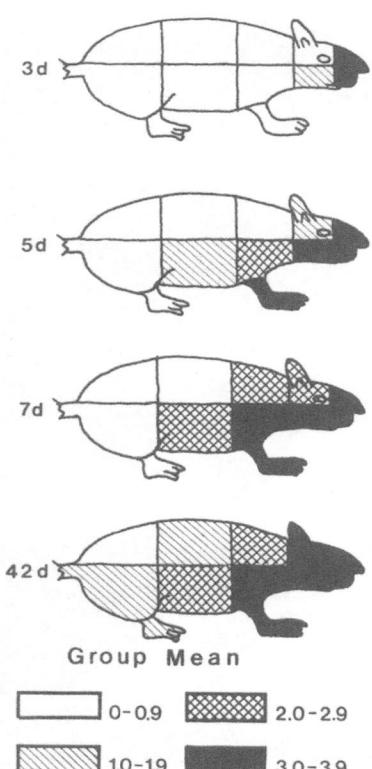

Group Mean

| | 0-0.9 | | 2.0-2.9 |
| | 1.0-1.9 | | 3.0-3.9 |

Figure 2. Somatotopic spread of zones of tactile sensitivity on the body surface contralateral to the 6-OHDA injection during the postoperative recovery period. Darker shadings (higher numbers) reflect more precise localization of a standard stimulus (von Frey hair exerting 4 g of force).

terminals throughout the forebrain, the extent of the impairment in somatosensory localization correlates significantly with the depletion of dopamine from the affected neostriatum (Marshall, 1979; Kozlowski and Marshall, 1981). In contrast, neither the depletion of dopamine from limbic or cortical regions nor the depletion of norepinephrine from any region analyzed reliably predicts the extent of the behavioral deficit. Thus, the loss of somatosensory localization appears to result quantitatively from damage to the neostriatal dopaminergic innervation.

Also, injections of 6-OHDA into the neostriatum that significantly decrease the dopaminergic innervation of this structure without damaging the innervation of limbic or cortical regions result in impairments in somatosensory localization. In contrast, 6-OHDA injections into the NAS, lateral septum, OT, or medial frontal cortex that damage the dopaminergic innervation of these sites do not (Marshall *et al.*, 1980).

To test more specifically the role of dopamine in this inattention, we determined whether the behavioral symptoms could be reversed by administration of the dopamine receptor agonist, apomorphine. When given systemically or by intracerebral injection into the neostriatum of rats made inattentive by prior 6-OHDA treatment, apomorphine reinstates somatosensory localization for up to 2 hr (Fig. 4; Marshall and Gotthelf, 1979; Marshall *et al.*, 1980). This apomorphine-induced reversal of the symptoms is prevented by prior administration of the dopamine receptor antagonist, haloperidol (Marshall and Gotthelf, 1979). The effectiveness of apomorphine in reversing the neglect suggests a tonic, modulatory role for dopamine within the neostriatum.

Finally, in an exciting recent report, Dunnett *et al.* (1981) have found that the ability of 6-OHDA-injected rats to localize stimuli can be reinstated by the implantation of fetal nigral dopamine-containing cells adjacent to the denervated neostriatum. These fetal implants send fluorescent axons into the denervated neostriatum, apparently providing a functional dopaminergic reinnervation.

5. NEURAL EVENTS CONTRIBUTING TO RECOVERY OF SENSORIMOTOR FUNCTIONS

5.1. Basal Gangliar Glucose Utilization

Although many rats with nigrostriatal injury recover the ability to localize somatosensory stimuli, not all do. When sacrificed several weeks postoperatively, those rats that recovered have reliably more dopamine remaining in their neostriata (8.3 ± 0.9% of control values) than do nonrecovering animals (2.8 ± 0.8%, $P < 0.01$; Kozlowski and Marshall, 1981). This difference suggests that the postoperative reinstatement of sensorimotor functions requires the survival of at least a small proportion (approximately 5%) of the normal population of

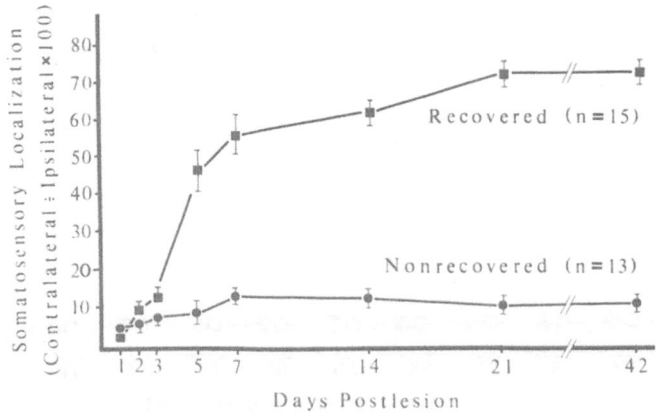

Figure 3. Quantification of the time course of recovery of contralateral somatosensory orientation for a group of rats that recovered this behavior postoperatively and another that did not. [Reprinted with permission from Kozlowski and Marshall (1983).]

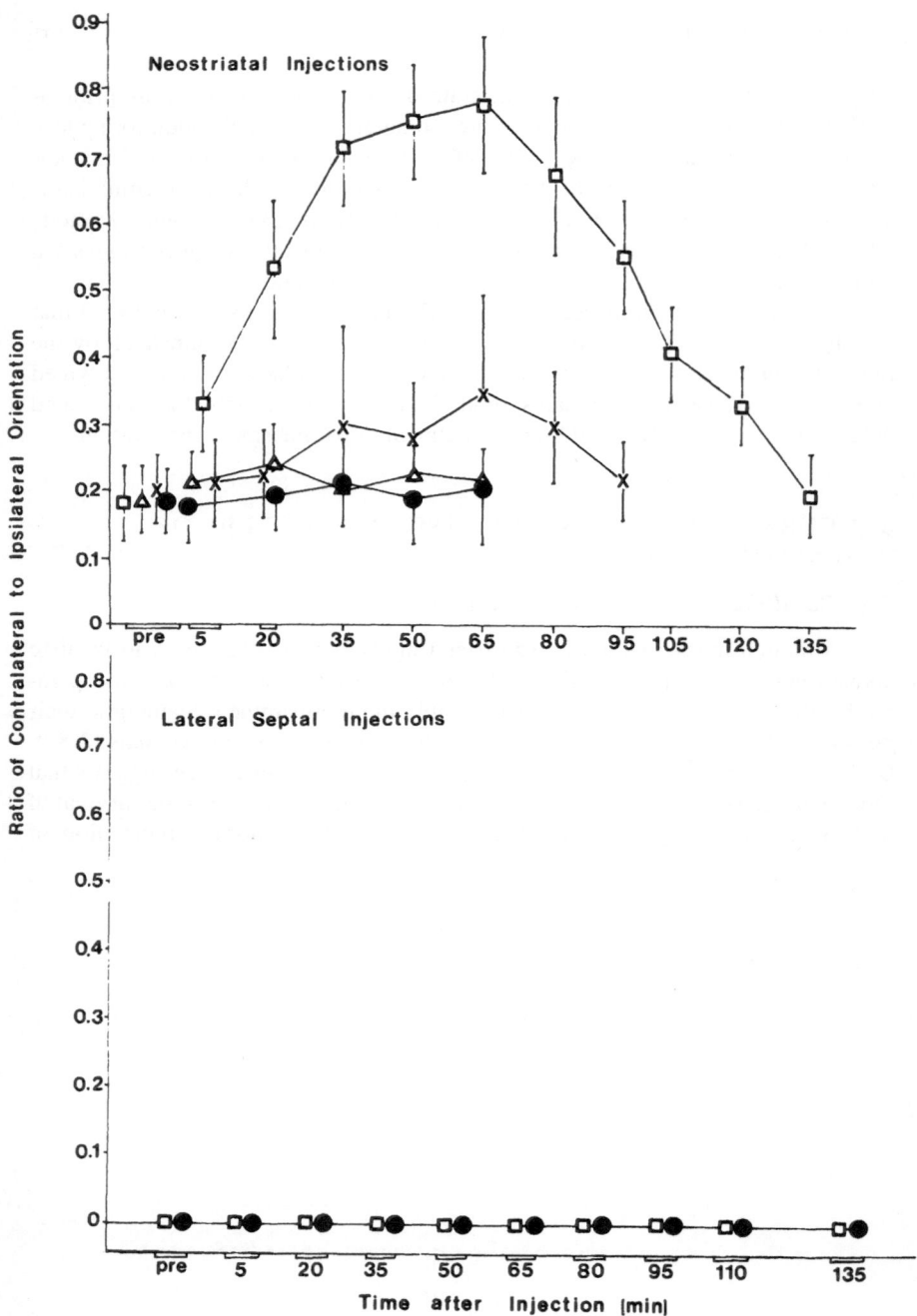

Figure 4. Upper left panel depicts the time course of reversal of contralateral somatosensory inattention following injection of 6 μg of apomorphine into the denervated neostriatum. No significant reversal of the symptoms occurred following injection of 3 μg of apomorphine or vehicle

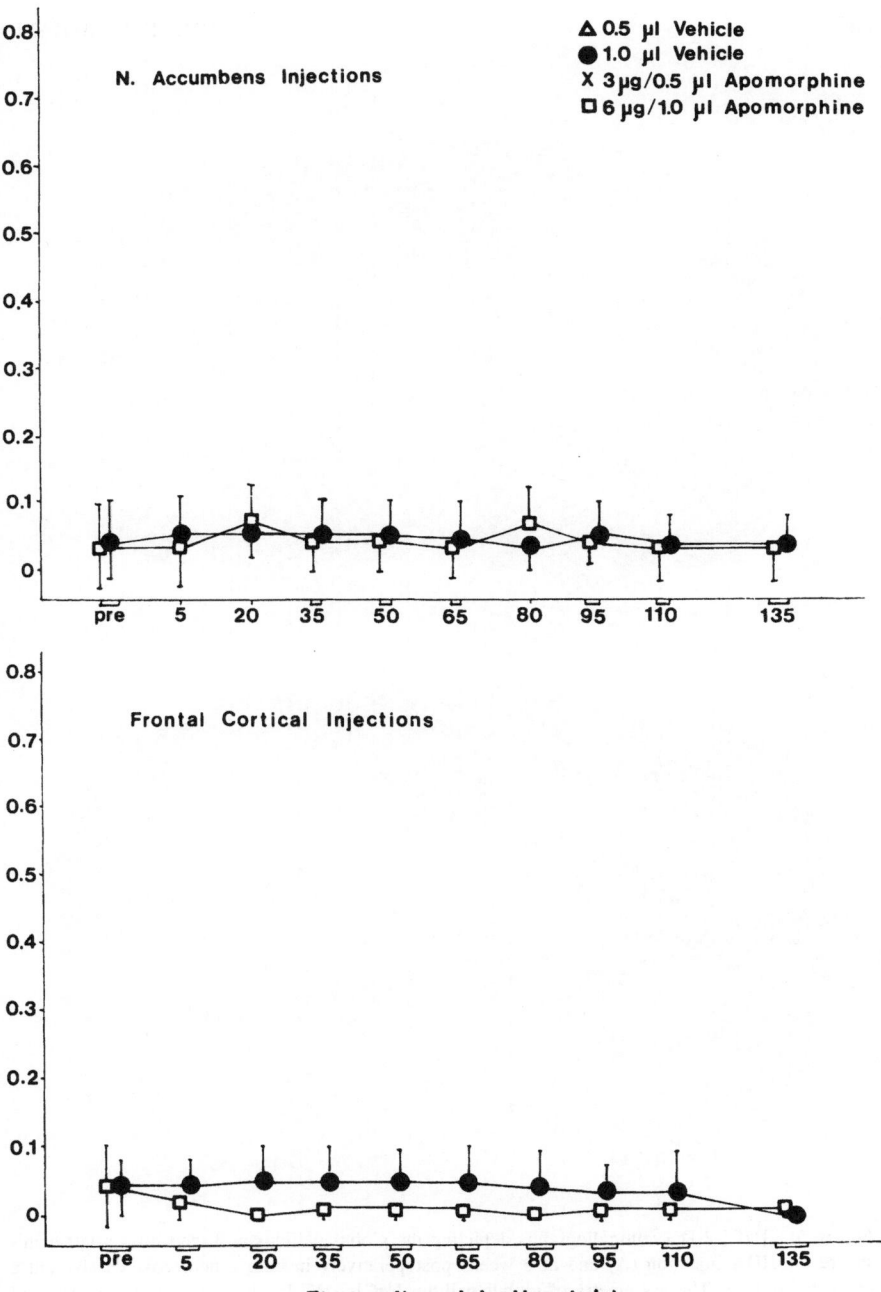

solutions. The remaining 3 panels indicate that no reversal of the inattention occurred after injections of 6 µg of apomorphine into the nucleus accumbens septi, lateral septum, or frontal cortex. [Reprinted with permission from Marshall *et al.* (1980).]

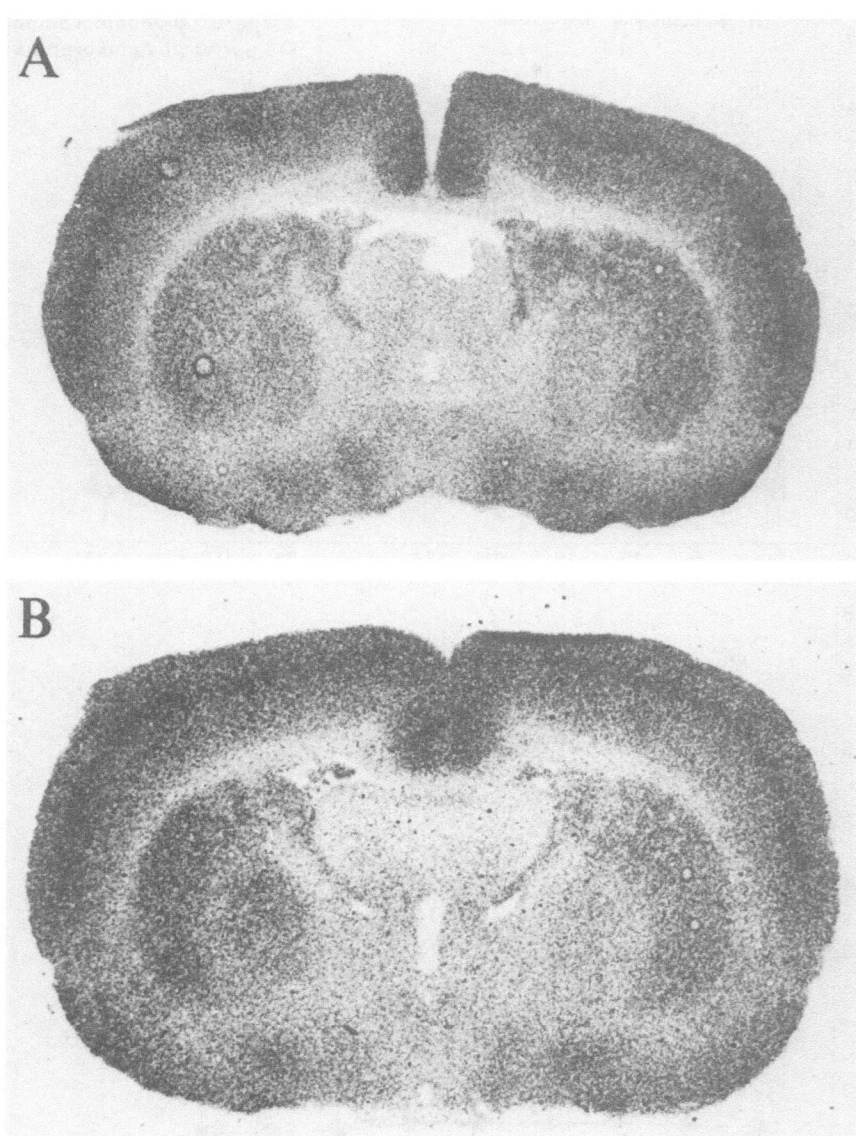

Figure 5.   [14C]-2-DG autoradiographs depicting the globus pallidus at 3 days after a left hemisphere 6-OHDA injection (A) and at 6 weeks postoperatively in both a nonrecovered (B) and a recovered (C) rat. The asymmetry of globus pallidus [14C]-2-DG labeling present at 3 days is still observed at 6 weeks in the nonrecovered animal but not in the recovered rat. [Reprinted with permission from Kozlowski and Marshall (1983).]

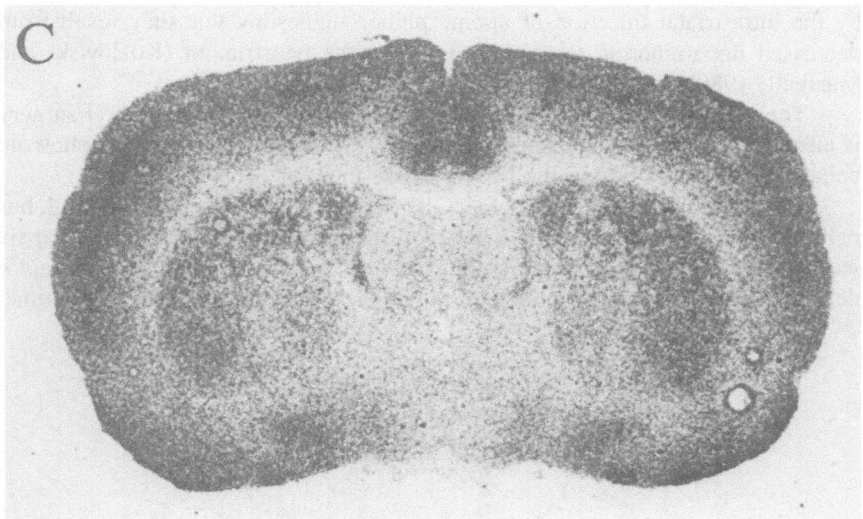

Figure 5 (continued)

dopaminergic nerve terminals in this structure. Perhaps recovery is mediated by synaptic changes occurring within this population of residual dopaminergic terminals.

Dr. Michael Kozlowski and I examined the contribution of the neostriatum to the recovery of sensorimotor functions using [^{14}C]-2-deoxyglucose (2-DG) autoradiography (Sokoloff *et al.*, 1977). Our goal was to determine the effect of nigrostriatal injury on [^{14}C]-2-DG incorporation into the neostriatum and related basal gangliar structures shortly after the 6-OHDA injection, at a time when somatosensory localization was greatly impaired. We then studied the brains of other rats killed six weeks after the injury. Some of these animals had recovered from their behavioral impairments during this six-week period whereas others had not. We determined whether the lesion-induced alterations in basal gangliar glucose utilization were reversed in either group during the postoperative survival period. Brain structures ipsilateral to the injury were compared with those of the contralateral control hemisphere.

Those animals that have a marked deficit in somatosensory localization at three days postoperatively show significant hemispheric asymmetries of [^{14}C]-2-DG utilization at this time. The uptake of this glucose analog is depressed ipsilaterally to the lesion in forebrain structures that normally receive a dense dopaminergic innervation (i.e., the neostriatum, NAS, OT, and central nucleus of the amygdala). In these same animals the ipsilateral uptake of [^{14}C]-2-DG is enhanced within structures that receive axonal projections *from* the neostriatum (i.e., globus pallidus, entopeduncular nucleus, and substantia nigra pars reticulata; Fig. 5). Most of these lesion-induced [^{14}C]-2-DG changes are reversed by the intrastriatal injection of apomorphine, suggesting that they result from decreased dopaminergic transmission within the neostriatum (Kozlowski and Marshall, 1980).

The pattern of altered [^{14}C]-2-DG uptake observed three days after surgery is also seen at six weeks postoperatively in 6-OHDA-injected rats that show no behavioral recovery during the postoperative interval.

However, the asymmetries of [^{14}C]-2-DG uptake in the neostriatum, globus pallidus (Fig. 5), and substantia nigra pars reticulata are no longer evident at six weeks postoperatively in animals that have recovered from their sensorimotor deficit. Moreover, the time course of the reversal of these [^{14}C]-2-DG asymme-

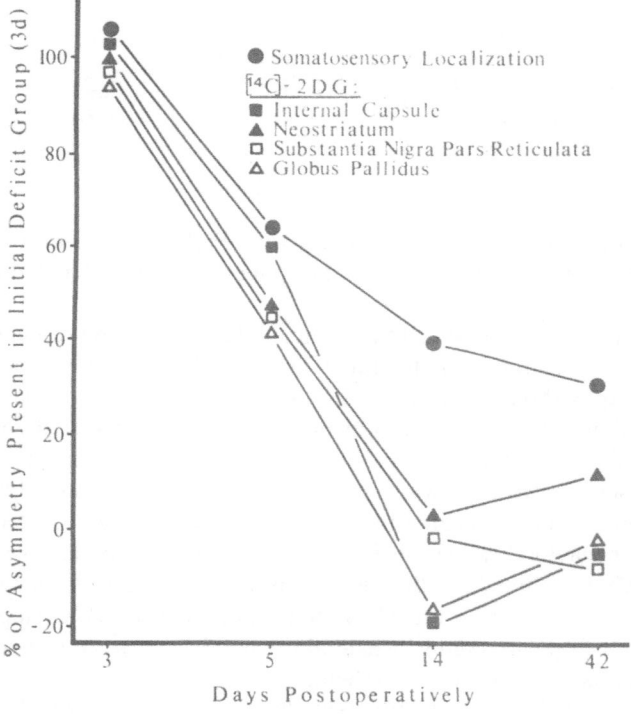

Figure 6. Time course for recovery of contralateral somatosensory localization compared with the time courses of the normalizations of [^{14}C]-2-DG uptake in the anterior neostriatum, globus pallidus, ventral internal capsule, and substantia nigra pars reticulata of recovering rats. Values represent the asymmetries in each of these measures at 3, 5, 14, and 42 days postoperatively as a percentage of the asymmetry at 3 days. Negative values represent a reversal of the direction of the asymmetry. The asymmetry in somatosensory localization is not fully reversed by 42 days because of residual impairments in orienting to touch of the most caudal body points. [Reprinted with permission from Kozlowski and Marshall (1983).]

tries in these basal gangliar structures is quite similar to that of the behavioral recovery (Fig. 6). In contrast, the lesion-induced decreases in [^{14}C]-2-DG uptake within the NAS and OT are just as extensive after six weeks of behavioral recovery as they are three days after the injury (Kozlowski and Marshall, 1983).

These findings indicate that the neostriatum plays a particularly important role in the recovery process. Because the intrastriatal injections of apomorphine in 6-OHDA-injected animals both restored contralateral somatosensory localization and reversed the neostriatal asymmetry of [^{14}C]-2-DG incorporation, we suggest that a reversal of the depressed glucose utilization of the denervated striatum may be a reliable index of the animals' recovery state, whether the recovery occurs spontaneously or is induced by a drug.

5.2. Compensatory Changes at Residual Neostriatal Dopaminergic Synapses and Recovery

The similar effects of spontaneous recovery and intrastriatal apomorphine administration on [^{14}C]-2-DG incorporation into the dopamine-depleted neostriatum has an additional implication. This similarity suggests that the spontaneous return of more normal somatosensory localization may be mediated by a time-dependent increase in the activity of dopaminergic synapses within this structure.

One possible explanation for this enhanced dopaminergic activity is that damaged dopaminergic axons regenerate processes from their proximal stumps and that these growing processes subsequently reinnervate neostriatal neurons. A second possibility is that dopamine-containing axons that survive the 6-OHDA injection grow collateral branches within the neostriatum that reinnervate the vacated target sites.

These explanations appear unlikely. First, although ascending catecholamine-containing axons do show a regenerative growth after their axons have been damaged by a 6-OHDA injection or mechanical transection, this growth response is limited to the parenchyma immediately surrounding the zone of injury (Katzman et al., 1971). In no instance have the ascending tracts been observed to reform. Second, if either regenerative growth or collateral sprouting of dopaminergic axons were to occur, these events should be marked by a postoperative return toward normal of several indices of dopaminergic nerve terminals, including the catecholamine histofluorescence of the neostriatum, its density of dopaminergic synaptic profiles, its dopamine content, its high affinity uptake of dopamine, or its content of the catecholamine synthetic enzyme, tyrosine hydroxylase. However, none of these anatomic, histochemical, or biochemical parameters are normalized in the neostriatum after the injury (Hökfelt and Ungerstedt, 1973; Breese and Traylor, 1971; Uretsky et al., 1971; Neve et al., 1983; Reis et al., 1978; Altar and Marshall, unpublished observations).

Thus, if regenerative or collateral sprouting of dopaminergic axons occurs after damage to this system, these events are of such small magnitude as to go undetected by several measures.

Assuming that these growth responses do not occur, what other events might increase the level of dopaminergic activity in the neostriatum after the injury? As noted previously, the recovery of somatosensory localization appears to require the survival of a small population (approximately 5%) of neostriatal dopaminergic terminals. In their studies of the feeding and drinking impairments of rats with brain dopamine depletions, Stricker and Zigmond (1976) observed a similar dependence of the recovery of these behaviors on the survival of a small critical proportion of these neurons. They suggested that the recovery of ingestive behaviors after injury results from a time-dependent increase in the level of activity at the residual synapses. Compensatory increases in the activity of these synapses could result from an enhanced synthesis and release (i.e., turnover) of dopamine by the remaining presynaptic elements, an elevated response of the postsynaptic neurons to released dopamine, or both (Stricker and Zigmond, 1976; Ungerstedt, 1971b). Presumably, rats with greater than 95% destruction to this neuronal population fail to recover from their impairments in somatosensory localization because these pre- and postsynaptic compensations are insufficient to restore the dopaminergic receptor activity of this structure to a level compatible with recovery.

Pharmacological experiments provide indirect support for the hypothesis that the recovery of somatosensory localization depends upon the continued functioning of the catecholaminergic neurons that survive the injury. After rats recover from their impairment in somatosensory localization following 6-OHDA injection, administration of a low dose of the dopamine receptor antagonist, spiroperidol (50 μg/kg, IP) reinstates the localization impairment for several hours. A similar reinstatement of the deficit occurred after administration of the catecholamine synthesis inhibitor, α-methylparatyrosine (50 or 100 mg/kg, IP). These results are compatible with the view that a continued synthesis, release, and receptor action of dopamine at the residual dopaminergic synapses is necessary for recovery (Marshall, 1979).

5.2.1. Changes in Dopamine Metabolism

Animals or humans with injury to the nigrostriatal dopaminergic projection have elevated ratios of the dopamine metabolites, homovanillic acid (HVA) and dihydroxyphenylacetic acid (DOPAC), to dopamine (Bernheimer and Hornykiewicz, 1962; Sharman et al., 1967; Hefti et al., 1980). These observations provided the first indication that after partial injury to this projection, the surviving dopaminergic nerve terminals increase their rate of transmitter turnover. This conclusion is supported by the finding that the rate of conversion of [^{3}H]tyrosine to [^{3}H]dopamine is increased within the neostriatum partially denervated of its dopaminergic afferents (Agid *et al.*, 1973).

Increases in dopamine turnover within surviving terminals could contribute to recovery by increasing the availability of dopamine at target cells. However, to our knowledge no attempt has been made to determine the time course of these turnover changes after injury. Because the recovery of somatosensory localization is not evident until 4–5 days postoperatively and is maximal by three weeks, Dr. Anthony Altar and I recently examined the extent to which dopamine turnover is elevated on days 5 and 18 after surgery (Fig. 7). At 18 days postoperatively, the ratios, DOPAC/dopamine and HVA/dopamine, are elevated in the partially denervated neostriatum relative to the control side. This apparent elevation of dopamine turnover is evident only for rats having greater than 90% depletions of neostriatal dopamine. The elevation of these ratios is greatest for those animals that sustained the most extensive damage to this projection (also Hefti *et al.*, 1980), suggesting that the extent of turnover increase in residual terminals depends upon the completeness of the denervation.

This elevation in the DOPAC/dopamine and HVA/dopamine ratios is also evident in the partially denervated neostriatum at five days postoperatively. Again, this apparent increase in dopamine turnover occurs only in rats sustaining greater than a 90% loss of neostriatal dopamine content. Furthermore, the quantitative functions that relate these elevated metabolite/dopamine ratios to the proportion of dopamine remaining in the neostriatum ipsilateral to the injury appear to be very similar for the two groups of animals (Fig. 7).

These results suggest that the elevation of neostriatal dopamine turnover occurs within five days after the injury (perhaps much sooner) and that no further increase occurs during the subsequent two weeks. This time course contrasts with that of the somatosensory recovery, which is incomplete by five days after injury (Fig. 3). Our findings indicate that although enhanced dopamine turnover may contribute to the behavioral recovery by helping to restore more normal levels of dopaminergic activity in the partially denervated neostriatum, other cellular events occurring later in the postoperative period must contribute to the later stages of the recovery.

5.2.2. Changes in Neostriatal Sensitivity to Dopamine

The existence of a postsynaptic supersensitivity of neostriatal cells to dopamine or dopamine agonist compounds after interruption of the dopaminergic afferents to this structure is indicated by behavioral (Ungerstedt, 1971c; Schoenfeld and Uretsky, 1972), electrophysiological (Feltz and de Champlain, 1972; Schultz and Ungerstedt, 1979), and neurochemical techniques (Creese *et al.*,

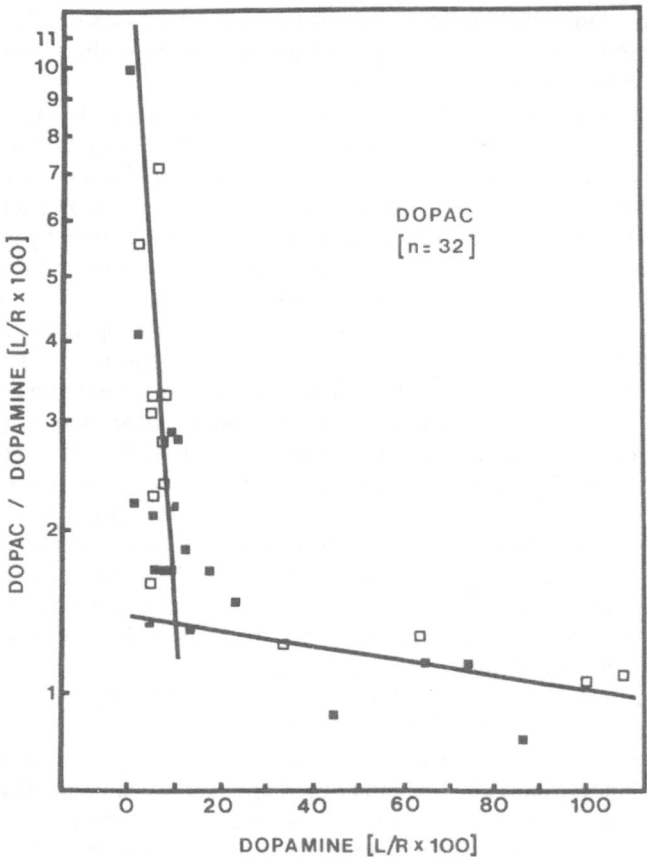

Figure 7. Scatter plots depicting the relationship of elevations of ipsilateral dopamine metabolite (DOPAC, HVA) concentrations to dopamine depletions in animals with unilateral 6-OHDA injections. The ratios of DOPAC/DA and HVA/DA in the ipsilateral neostriatum are expressed on the ordinate as a percent of that in the contralateral side for each animal. The abcissa depicts the percent

1977; Mishra *et al.*, 1974; Fibiger and Grewaal, 1974). This enhanced sensitivity may contribute to the improvement of behavioral functions after nigrostriatal injury by increasing the response of neostriatal cells to the available dopamine.

In denervated skeletal and smooth muscles, where the basis of this supersensitivity has been studied extensively, at least three cellular events appear to contribute to this supersensitivity:

1. In skeletal muscle, denervation alters the distribution of acetylcholine receptors on the membrane of the muscle cell, such that the density of

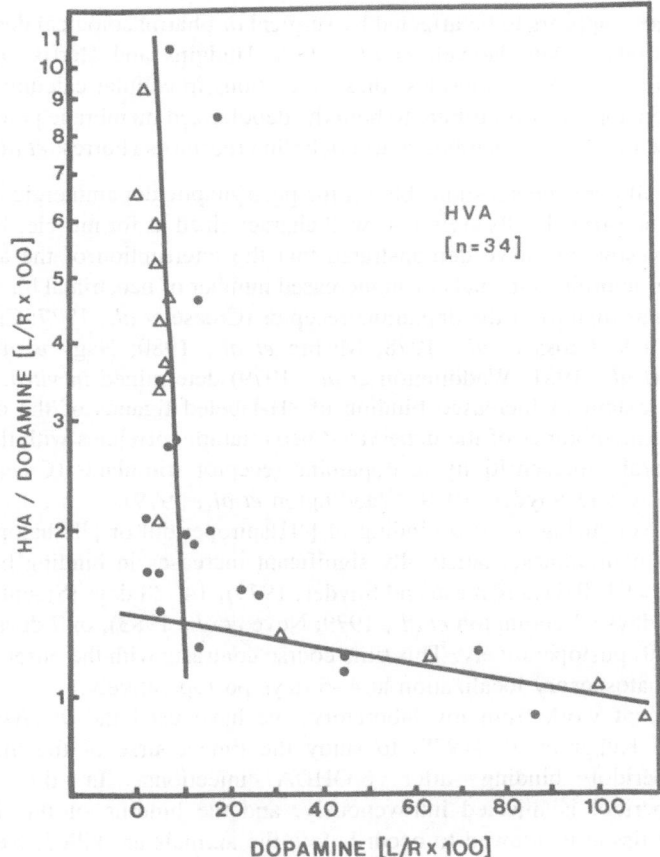

of dopamine in the ipsilateral (left) neostriatum as a percent of that on the contralateral (right) side. In each frame, open symbols represent rats killed at 18 days postoperatively and closed symbols represent those killed at 5 days after surgery.

these receptors outside of the end-plate region increases dramatically (Axelsson and Thesleff, 1959).

2. In both smooth and skeletal muscle, denervation leads to a depolarization of the muscle membrane, rendering the cell more likely to contract in response to depolarizing chemical stimuli (Fleming and Westfall, 1975; Locke and Solomon, 1967). In smooth muscle, this depolarization appears to be mediated by a reduced activity of an electrogenic sodium pump (Gerthoffer et al., 1979).

3. In both smooth and skeletal muscle, the intracellular distribution of

Ca^{2+} appears to be affected by surgical or pharmacological denervation (Brody, 1966; Howell et al., 1966; Hudgins and Harris, 1970). In cultured skeletal muscles, these alterations in cellular calcium distribution appear to contribute to both the depolarized membrane potential and the increase in membrane acetylcholine receptors (Forrest et al., 1981).

The cellular events responsible for the postsynaptic dopaminergic supersensitivity of neostriatal cells are not as well characterized as for muscle. However, several investigators have demonstrated that the interruption of the ascending dopaminergic projection leads to an increased number of neostriatal binding sites for particular ligands of the dopamine receptor (Creese et al., 1977; Creese and Snyder, 1979; Cross et al., 1978; Mishra et al., 1980; Nagy et al., 1979; Staunton et al., 1981; Waddington et al., 1979) determined in vitro. Furthermore, the extent of increased binding of ^{3}H-labeled ligands of the dopamine receptor to membranes of the denervated neostriatum correlates with the degree of behavioral supersensitivity to dopamine receptor stimulants (Creese et al., 1977; Creese and Snyder, 1979; Waddington et al., 1979).

However, using in vitro binding of [^{3}H]spiroperidol or [^{3}H]haloperidol to neostriatal membranes, statistically significant increases in binding have been found only at 120 days (Creese and Snyder, 1979), 14–28 days (Staunton et al., 1981), 21 days (Waddington et al., 1979; Neve et al., 1983), or 7 days (Mishra et al., 1980) postoperatively. This time course contrasts with the onset of recovery of somatosensory localization at 4–5 days postoperatively.

In recent work from my laboratory, we have used the in vivo binding method of Kuhar et al. (1978) to study the time course of the increase in [^{3}H]spiroperidol binding after 6-OHDA injections. In this method, [^{3}H]spiroperidol is injected intravenously, and the binding of this ligand to neostriatal tissue is allowed to occur before the animals are killed. Neve et al. (1983) find that [^{3}H]spiroperidol binding to the denervated neostriatum is significantly increased by four days postoperatively. This elevation of binding increases linearly in magnitude during the first postoperative month. Thus, the in vivo binding method appears to be more sensitive to these early effects of denervation than are the in vitro methods. The time course of these in vivo binding changes is consistent with the view that an increase in at least one class of dopamine receptors contributes to the recovery of somatosensory localization.

5.3. Recovery from Nigrostriatal Injury in Senescence

To explore further the role of these receptor changes in the recovery process, the effect of nigrostriatal 6-OHDA injection on somatosensory localization and [^{3}H]spiroperidol binding was compared in aged and young adult rats. Old age has important consequences for the structure and neurochemistry of the basal ganglia, and particularly for the nigrostriatal dopaminergic projection (reviewed by Finch et al., 1981). These age-dependent changes in the basal ganglia appear to impair the motor functions of the elderly. Administration of L-dopa or the dopamine receptor stimulant, apomorphine, to aged rats restores their swimming abilities to that of young adult controls (Marshall and Berrios, 1979).

In advanced age, some forms of neuronal plasticity observed in response to nervous system injury are slowed or diminished in magnitude (Scheff et al., 1980; Hoff et al., 1981). Because of these findings my associates and I were interested in determining whether the recovery of somatosensory localization after nigrostriatal injury is slowed or decreased in extent during old age.

Different doses of 6-OHDA (2,4, or 6 μg) were injected unilaterally along the course of the nigrostriatal projection in young adult (4–5 months) and aged (27–28 months) rats to produce a range of depletions of neostriatal dopamine. The extent of recovery of contralateral somatosensory localization was determined during the first month postoperatively, after which the catecholamine content of each neostriatum was measured. Some of the rats with the most extensive behavioral impairments were also tested for behavioral supersensitivity by measuring apomorphine-induced rotational behavior (Ungerstedt, 1971c) and for in vivo [^{3}H]spiroperidol binding.

Aged rats are very sensitive to the behavioral effects of 6-OHDA. The aged animals given injections of 2 μg of 6-OHDA, for example, show much less recovery of somatosensory localization than do young adult rats given the same dose of neurotoxin (35 versus 86% recovery, $P < 0.01$). This age difference in recovery is explained by the finding that the aged animals sustain a more extensive depletion (85% loss) of neostriatal dopamine than do young adults (30%) at this 6-OHDA dose ($P < 0.002$). When the results for animals given each of the three 6-OHDA dose levels are combined, no difference can be observed between the recovery of old and young adult rats that sustained equivalent neostriatal dopamine depletions (Fig. 8). Thus, the ability of aged rats to recover from the behavioral deficits induced by a particular level of brain dopamine depletion appears equal to that of young adults. The basis for the enhanced neurotoxicity of 6-OHDA in aged rats is currently under investigation in our laboratory, but age differences in the placement of the 6-OHDA injection cannula or in the extent of the "nonspecific" tissue damage resulting from the 6-OHDA injection do not account for this age change (Marshall et al., 1983).

Not only do aged and young adult rats recover their sensorimotor functions equivalently after nigrostriatal injury, they show similar degrees of behavioral supersensitivity to the dopamine receptor stimulant, apomorphine. When tested 5–6 weeks postoperatively, both groups showed vigorous contralateral rotation after administration of this drug (0.25 mg/kg, IP).

These behavioral and pharmacological findings are paralleled by our investigations of in vivo [^{3}H]spiroperidol binding. Both young adult and aged rats

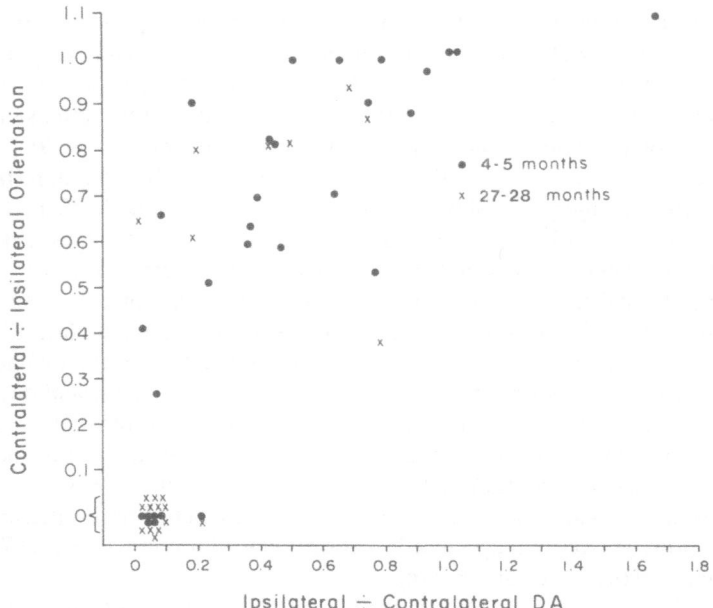

Figure 8. Scatter plot depicting the extent of recovery of contralateral somatosensory localization for young and aged rats during the 28-day postoperative examination period as a function of the dopamine content remaining in the neostriatum ipsilateral to the 6-OHDA injection. The ordinate depicts the ratio of scores for contralateral to ipsilateral orientation, whereas the abcissa depicts the ratio of the dopamine content for the ipsilateral to contralateral neostriata. The distribution of points for individual young adult and aged rats overlaps extensively. [Reprinted with permission from Marshall *et al.* (1983).]

show significant elevations in the binding of this ligand to the denervated striatum relative to that of contralateral hemisphere (23% and 29% increases respectively). These results are in agreement with those of Joseph *et al.* (1981), using *in vitro* binding of [³H]neuroleptics.

Our findings suggest that in spite of the marked alterations in the functioning of nigrostriatal dopaminergic synapses in advanced age, the capacity for plasticity in this system after injury is remarkably intact. In addition to the capacity for receptor proliferation demonstrated here, the neostriatal dopaminergic terminals of the aged rat increase their rate of neurotransmitter turnover in response to neuroleptic treatment (Severson *et al.*, 1981), as do young adult animals. Thus, both presynaptic and postsynaptic forms of dopaminergic synaptic plasticity occur in aged rats, just as in young adult animals.

5.4. Lithium Effects on Recovery

To test further the contribution of dopamine receptor proliferation to the recovery from nigrostriatal injury, Michael Kozlowski, Kim Neve, Jonathan Grisham, and I investigated the effect of chronic lithium administration on this recovery. Pert *et al.* (1978) found that treatment of rats with lithium during a period of chronic haloperidol administration blocked both the proliferation of [³H]spiroperidol binding sites and the behavioral supersensitivity to dopamine agonists that occurs following chronic haloperidol treatment alone. Given this result, lithium treatment appeared to provide an interesting means of testing the role of [³H]spiroperidol binding site proliferation in this recovery.

Rats that are given unadulterated tap water for four weeks postoperatively show the anticipated improvement in contralateral somatosensory localization during this time (Fig. 9). When this group of animals is switched to lithium water (1.25 g/liter) between weeks 4 and 8, they maintain their recovered status.

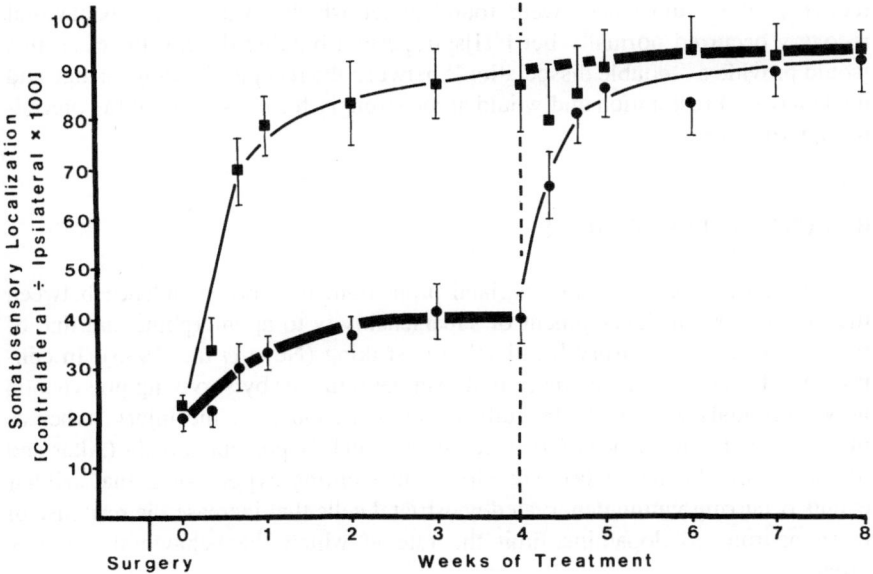

Figure 9. Lithium effects on the recovery of contralateral somatosensory localization after unilateral ventral tegmental 6-OHDA injections. The broad line indicates treatment with lithium-containing water, whereas the narrow line shows administration of unadulterated tap water. Reversal of the treatment conditions occurred at 4 weeks (broken vertical line). The results are based on 11 rats that started the experiment on lithium and 10 that started on unadulterated water, all of which recovered somatosensory localization by 8 weeks postoperatively.

However, rats that are given lithium in their drinking water for the first four weeks postoperatively show little recovery of somatosensory localization during this time, but they recover rapidly when allowed access to unadulterated water starting four weeks after surgery.

Surprisingly, however, lithium treatment does not prevent the increase in *in vivo* [³H]spiroperidol binding in the denervated neostriatum, relative to the contralateral hemisphere. At four weeks postoperatively, both lithium-treated and control rats show a 22% greater binding of [³H]spiroperidol in the denervated than in the intact neostriatum (Kozlowski *et al.*, 1981).

These findings appear to dissociate the behavioral recovery after nigrostriatal injury from the lesion-induced proliferation of neostriatal [³H]spiroperidol binding sites. One possible explanation of the results is that lithium may interfere with cellular events occurring subsequent to the binding of dopamine ligands to the receptor sites. If so, then a proliferation of the binding sites in drug-free rats might contribute to the recovery, whereas this proliferation might be insufficient for the recovery to occur in lithium-treated rats. Alternatively, it may be that a proliferation of the [³H]spiroperidol binding sites does not normally contribute to recovery. If circumstances were found under which postoperative behavioral recovery occurred normally but [³H]spiroperidol binding did not increase, this would provide a "double dissociation" between the receptor binding change and the behavioral restoration and would argue strongly for the second of the preceding alternatives.

6. FUTURE DIRECTIONS

After damage to the nigrostriatal projection, the correspondence between the time course of development of supersensitivity to apomorphine and that of recovery of somatosensory localization is striking (Neve *et al.*, 1983). In contrast, the lesion-induced increase in dopamine turnover by surviving presynaptic nerve terminals appears to be fully developed soon after the injury, whereas much of the recovery occurs between days 5 and 21 postoperatively (Altar and Marshall, unpublished data). Therefore, our working hypothesis is that cellular events occurring within denervated neostriatal cells that increase the response of these neurons to dopamine limit the rate at which the behavioral recovery proceeds.

The increase in neostriatal dopamine binding sites labeled by [³H]spiroperidol occurs with a time course consistent with that of recovery, and this increase is also evident in the brains of senescent recovering animals. These results are consistent with the view that an increase in striatal dopamine receptor binding mediates the observed recovery. However, our observations concerning lithium's effect on recovery suggest that the denervation-induced increase in neostriatal [^{3}H]spiroperidol is, at the least, not a sufficient condition for the recovery. What other cellular events occur within the denervated neostriatum that may contribute to this recovery?

One possibility is that changes in neostriatal dopamine receptors do mediate the recovery of function observed, but that the population of such receptors relevant to this recovery is not labeled by [^{3}H]spiroperidol. Whereas the identification and classification of dopamine receptors in the neostriatum has been a controversial topic, there now appear to be three, and possibly four, binding sites (Seeman, 1980; Sokoloff et al., 1980). These sites differ in their affinity for different ligands, their cellular localizations in the neostriatum, and other characteristics. The judicious use of other ligands for dopamine receptors may elucidate their contribution to this recovery sequence.

In addition, events occurring subsequent to the dopamine–receptor interaction may contribute to the increased response of neostriatal cells to this transmitter. For example, one hypothesis of how dopamine inhibits the activity of neostriatal neurons is that the transmitter hyperpolarizes the membrane by activating an electrogenic sodium pump (Yarbrough, 1976). Kamata et al. (1980) found that dopamine results in a greater stimulation of Na^+,K^+-ATPase in the neostriatum ipsilateral to a nigrostriatal 6-OHDA injection than on the contralateral side (however, see van der Krogt and Belfroid, 1980).

An additional possibility is that the supersensitivity of denervated neostriatal neurons to dopamine involves a redistribution of intracellular calcium. Chronic treatment of rats with dopamine-receptor-blocking drugs [e.g., haloperidol or (+)-butaclamol] induces a behavioral supersensitivity to the dopamine agonist, apomorphine, and an increased neostriatal membrane content of a calcium-dependent activator of adenylate cyclase (Gnegy et al., 1970a,b). In neuroleptic-treated supersensitive animals the release of this calcium-dependent protein from the neostriatal membrane is also reduced, which may reflect a change in calcium sequestration in neostriatal neurons deprived of their dopaminergic receptor activity (Gnegy and Lau, 1980).

7. CONCLUSIONS

The recovery of somatosensory localization that occurs after damage to the nigrostriatal dopaminergic projection appears to be a consequence of adaptations occurring at synapses of this system that survive the injury. These forms of synaptic plasticity lead to a normalization of the rate of glucose utilization within the partially denervated neostriatum.

These findings suggest that a "vicarious functioning" within neural structures extrinsic to the original injury is not responsible for this instance of recovery. If the destruction of this neural system exceeds a critical limit (i.e., if neostriatal dopamine is depleted by 95% or more), behavioral recovery does not occur and neostriatal glucose utilization rate remains depressed.

An increased rate of dopamine synthesis and release within the surviving axon terminals likely contributes to this recovery by increasing the concentration of dopamine available to interact with receptors on neostriatal neurons. However, this form of presynaptic plasticity is insufficient for behavioral restoration, and the rate at which the recovery of somatosensory localization proceeds is likely to depend upon cellular events occurring within neostriatal neurons denervated of much of their dopaminergic innervation. An understanding of the postsynaptic contributions to this recovery sequence requires a comprehensive analysis of the cellular changes occurring within denervated neostriatal cells.

REFERENCES

Acheson, A. L., Zigmond, M. J., and Stricker, E. M., 1980, Compensatory increase in tyrosine hydroxylase in rat brain after intraventricular injections of 6-hydroxydopamine, *Science* **207**:537–540.

Agid, Y., Javoy, F., and Glowinski, J., 1973, Hyperactivity of DA neurons after partial destruction of nigro-striatal DA system, *Nature New Biol.* **245**:150–151.

Axelsson, J., and Thesleff, S., 1959, A study of supersensitivity in denervated mammalian skeletal muscle, *J. Physiol.* **147**:178–193.

Bernheimer, H., and Hornykiewicz, O., 1962, Das Verhalten einiger Enzyme im Gehirn normaler und Parkinson-Kranker Menschen, *Naunyn-Schmiedeberg's Arch. Exp. Pathol.* **243**:295–296.

Breese, G. R., and Traylor, T. D., 1971, Depletion of brain noradrenaline and dopamine by 6-hydroxydopamine, *Br. J. Pharmacol.* **42**:88–99.

Brody, I. A., 1966, Relaxing factor in denervated muscle: A possible explanation for fibrillations, *Am. J. Physiol.* **211**:1277–1280.

Cotman, C. (ed.), 1978, *Neuronal Plasticity*, Raven Press, New York.

Creese, I., and Snyder, S. H., 1979, Nigrostriatal lesions enhance striatal ^{3}H-apomorphine and ^{3}H-spiroperidol binding, *Eur. J. Pharmacol.* **56**:277–281.

Creese, I., Burt, D. R., and Snyder, S. H., 1977, Dopamine receptor binding enhancement accompanies lesion-induced behavioral supersensitivity, *Science* **197**:596–598.

Cross, A. J., Longden, A., Owen, F., Poulter, M., and Waddington, J. L., 1978, Inter-relationship between behavioral and neurochemical indices of supersensitivity in dopaminergic neurones, *Br. J. Pharmacol.* **63**:383P.

Dunnett, S. B., Björklund, A., Stenevi, U., and Iversen, S. D., 1981, Grafts of embryonic substantia nigra reinnervating the ventrolateral striatum ameliorate sensorimotor impairments and akinesia in rats with 6-OHDA lesions of the nigrostriatal pathway, *Brain Res.* **229**:209–217.

Fallon, J. H., and Moore, R. Y., 1978, Catecholamine innervation of the basal forebrain. IV. Topography of the dopamine projection to the basal forebrain and neostriatum, *J. Comp. Neurol.* **180**:545–580.

Feltz, P., and de Champlain, J., 1972, Enhanced sensitivity of caudate neurones to microiontophoretic injections of dopamine in 6-hydroxydopamine treated rats, *Brain Res.* **43**:601–605.

Fibiger, H. C., and Grewaal, D. S., 1974, Neurochemical evidence for denervation supersensitivity: The effect of unilateral substantia nigra lesions on apomorphine-induced increases in neostriatal acetylcholine levels, *Life Sci.* **15**:57–63.

Finch, C. E., Randall, P. K., and Marshall, J. F., 1981, Aging and basal gangliar functions, *Annu. Rev. Gerontol. Geriatr.* **2**:49–87.

Fleming, W. W., and Westfall, D. P., 1975, Altered resting membrane potential in the supersensitive vas deferens of the guinea pig, *J. Pharmacol. Exp. Ther.* **192**:381–389.

Forrest, J. W., Mills, R. G., Bray, J. J., and Hubbard, J. I., 1981, Calcium-dependent regulation of the membrane potential and extrajunctional acetylcholine receptors of rat skeletal muscle, *Neuroscience* **6**:741–749.

Gerthoffer, W. T., Fedan, J. S., Westfall, D. P., Goto, K., and Fleming, W. W., 1979, Involvement of the sodium-potassium pump in the mechanism of postjunctional supersensitivity of the vas deferens of the guinea pig, *J. Pharmacol. Exp. Ther.* **210**:27–36.

Gnegy, M. E., and Lau, Y. S., 1980, Effects of chronic and acute treatment of antipsychotic drugs on calmodulin release from rat striatal membranes, *Neuropharmacology* **19**:319–323.

Gnegy, M. E., Lucchelli, A., and Costa, E., 1977a, Correlation between drug-induced supersensitivity of dopamine dependent striatal mechanisms and the increase in striatal content of the Ca^{2+} regulated protein activator of cAMP phosphodiesterase, *Naunyn-Schmiedeberg's Arch. Pharmacol.* **301**:121–127.

Gnegy, M., Uzunov, P., and Costa, E., 1977b, Participation of an endogenous Ca^{++}-binding protein activator in the development of drug-induced supersensitivity of striatal dopamine receptors, *J. Pharmacol. Exp. Ther.* **202**:558–564.

Goldberger, M. E., and Murray, M., 1978, Recovery of movement and axonal sprouting may obey some of the same laws, in: *Neuronal Plasticity* (C. W. Cotman, ed.), Raven Press, New York, pp. 73–96.

Guth, L., 1974, Axonal regeneration and functional plasticity in the central nervous system, *Exp. Neurol.* **45**:606–654.

Hefti, F., Malamed, E., and Wurtman, R. J., 1980, Partial lesions of the nigrostriatal system in rat brain: Biochemical characterization, *Brain Res.* **195**:123–137.

Hoff, S. F., Scheff, S. W., Kwan, A. Y., and Cotman, C. W., 1981, A new type of lesion-induced synaptogenesis: II. The effect of aging on synaptic turnover in non-denervated zones, *Brain Res.* **222**:15–27.

Hökfelt, T., and Ungerstedt, U., 1973, Specificity of 6-hydroxydopamine-induced degeneration of central monoamine neurones: An electron and fluorescence microscopic study with special reference to intracerebral injection on the nigro-striatal dopamine system, *Brain Res.* **60**:269–297.

Howell, J. N., Fairhurst, A. S., and Jenden, D. J., 1966, Alterations of the calcium accumulating ability of striated muscle following denervation, *Life Sci.* **5**:439–446.

Hudgins, P. M., and Harris, T. M., 1970, Further studies on the effects of reserpine pretreatment on rabbit aorta: Calcium and histologic changes, *J. Pharmacol. Exp. Ther.* **175**:609–618.

Joseph, J. A., Filburn, C. R., and Roth, G. S., 1981, Development of dopamine receptor denervation supersensitivity in the neostriatum of the senescent rat, *Life Sci.* **29**:575–584.

Kamata, K., Kamata, M., Ino, H., and Kumeyama, T., 1980, Change in catecholamine-sensitive Na^+,K^+-ATPase activity in the rat striatum microsomes following electrolytic or 6-hydroxydopamine-induced lesions of dopaminergic neurons, *Jpn. J. Pharmacol.* **30**:401–404.

Katzman, R., Björklund, A., Owman, Ch., Stenevi, U., and West, K. A., 1971, Evidence for regenerative axon sprouting of central catecholamine neurons in the rat mesencephalon following electrolytic lesions, *Brain Res.* **25**:579–596.

Kozlowski, M. R., and Marshall, J. F., 1980, Plasticity of ^{14}C-2-deoxy-D-glucose incorporation into neostriatum and related structures in response to dopamine neuron damage and apomorphine replacement, *Brain Res.* **197**:167–183.

Kozlowski, M. R., and Marshall, J. F., 1981, Plasticity of neostriatal metabolic activity and recovery from nigrostriatal injury, *Exp. Neurol.* **74**:318–323.

Kozlowski, M. R., and Marshall, J. F., 1983, Recovery of function and basal ganglia ^{14}C-2-deoxyglucose uptake after nigrostriatal injury, *Brain Res.* **259**:237–248.

Kozlowski, M. R., Grisham, J., and Marshall, J. F., 1981, Chronic lithium treatment attenuates recovery following damage to the mesotelencephalic dopamine system, *Neurosci. Abstr.* **7**:627.

Kuhar, M. J., Murrin, L. C., Malouf, A. T., and Klemm, N., 1978, Dopamine receptor binding *in vivo:* The feasibility of autoradiographic studies, *Life Sci.* **22**:203–210.

Lindvall, O., and Björklund, A. N., 1974, The organization of the ascending catecholamine neuron systems in the rat brain as revealed by the glyoxylic acid fluorescence method, *Acta Physiol. Scand. (Suppl.)* **412**:1–48.

Lindvall, O., and Björklund, A., 1978, Anatomy of the dopaminergic neuron systems in the rat brain, in: *Advances in Biochemical Psychopharmacology,* Volume 19 (P. J. Roberts *et al.*, eds), Raven Press, New York, pp. 1–23.

Liu, C. N., and Chambers, W. W., 1958, Intraspinal sprouting of dorsal root axons, *Arch. Neurol. Psychiatr.* **79**:46–61.

Locke, S., and Solomon, H. C., 1967, Relation of resting potential of rat gastrocnemius and soleus muscles to innervation, activity, and the Na-K pump, *J. Exp. Zool.* **166**:377–386.

Loesche, J., and Steward, O., 1977, Behavioral correlates of denervation and reinnervation of the hippocampal formation of the rat: Recovery of alternation performance following unilateral entorhinal cortex lesions, *Brain Res. Bull.* **2**:31–39.

Low, W. C., Dunnett, S. B., Bunch, S. T., Thomas, S. R., Lewis, P. R., Iversen, S. D., Björklund, A., and Stenevi, U., 1981, Restoration of synaptic and behavioral function with embryonic transplants of cholinergic neurons, *Neurosci. Abstr.* **7**:259.

Marshall, J. F., 1979, Somatosensory inattention after dopamine depleting intracerebral 6-OH-DA injections: Spontaneous recovery and pharmacological control, *Brain Res.* **177**:311–324.

Marshall, J. F., 1980, Basal ganglia dopaminergic control of sensorimotor functions related to motivated behavior, in: *Neural Mechanisms of Goal-Directed Behavior and Learning* (R. F. Thompson, L. H. Hicks, and V. B. Shvyrkov, eds.), Academic Press, New York, pp. 167–176.

Marshall, J. F., and Berrios, N., 1979, Movement disorders of aged rats: Reversal by dopamine receptor stimulation, *Science* **206**:477–479.

Marshall, J. F., and Gotthelf, T., 1979, Sensory inattention in rats with 6-hydroxydopamine-induced degeneration of ascending dopaminergic neurons: Apomorphine-induced reversal of deficits, *Exp. Neurol.* **65**:398–411.

Marshall, J. F., and Teitelbaum, P., 1974, Further analysis of sensory inattention following lateral hypothalamic damage in rats, *J. Comp. Physiol. Psychol.* **86**:375–395.

Marshall, J. F., Turner, B. H., and Teitelbaum, P., 1971, Sensory neglect produced by lateral hypothalamic damage, *Science* **174**:523–525.

Marshall, J. F., Richardson, J. S., and Teitelbaum, P., 1974, Nigrostriatal bundle damage and the lateral hypothalamic syndrome, *J. Comp. Physiol. Psychol.* **87**:808–830.

Marshall, J. F., Berrios, N., and Sawyer, S., 1980, Neostriatal dopamine and sensory inattention, *J. Comp. Physiol. Psychol.* **94**:833–846.

Marshall, J. F., Drew, M. C. and Neve, K. A., 1983, Recovery of function after mesotelencephalic dopaminergic injury in senescence, *Brain Res.* **259**:249–260.

Millar, J., Basbaum, A. I., and Wall, P. D., 1976, Restructuring of the somatotopic map and appearance of abnormal neuronal activity in the gracile nucleus after partial deafferentation, *Exp. Neurol.* **50**:658–672.

Mishra, R. K., Gardner, E. L., Katzman, R., and Makman, M. H., 1974, Enhancement of dopamine-stimulated adenylate cyclase activity in rat caudate after lesions in substantia nigra: Evidence for denervation supersensitivity, *Proc. Natl. Acad. Sci. U.S.A.* **71**:3883–3887.

Mishra, R. K., Marshall, A. M., and Varmuza, S. L., 1980, Supersensitivity in rat caudate nucleus: Effects of 6-hydroxydopamine on the time course of dopamine receptor and cyclic AMP changes, *Brain Res.* **200**:47–57.

Murray, M., and Goldberger, M. E., 1974, Restitution of function and collateral sprouting in the cat spinal cord: The partially hemisected animal, *J. Comp. Neurol.* **158**:19–36.

Nagy, J. I., Lee, T., Seeman, P., and Fibiger, H. C., 1979, Direct evidence for presynaptic and postsynaptic dopamine receptors in brain, *Nature* **277**:93–96.

Nelson, D. L., Herbert, A., Bourgoin, S., Glowinski, J., and Hamon, M., 1978, Characteristics of central 5-HT receptors and their adaptive changes following intracerebral 5,7-dihydroxytryptamine administration in the rat, *Mol. Pharmacol.* **14**:983–995.

Neve, K. A., Kozlowski, M. R., and Marshall, J., 1983, Plasticity of neostriatal dopamine receptors after nigrostriatal injury: Relationship to recovery of sensorimotor functions and behavioral supersensitivity, *Brain Res.* **244**:33–44.

Nygren, L.-G., Olson, L., and Seiger, Å., 1971, Regeneration of monoamine-containing axons in the developing and adult spinal cord of the rat following intraspinal 6-OH-dopamine injections or transections, *Histochemistry*, **28**:1–15.

Nygren, L.-G., Fuxe, K., Jonsson, G., and Olson, L., 1974, Functional regeneration of 5-hydroxytryptamine nerve terminals in the rat spinal cord following 5,6-dihydroxytryptamine induced degeneration, *Brain Res.* **78**:377–394.

Pert, A., Rosenblatt, J. E., Sivit, C., Pert, C. B., and Bunney, W. E., Jr., 1978, Long-term treatment with lithium prevents the development of dopamine receptor supersensitivity, *Science* **201**:171–173.

Ramon y Cajal, S., 1928, *Degeneration and Regeneration of the Nervous System* (R. M. May, trans. and ed.), Oxford, London.

Reis, D. J., Gilad, G., Pickel, V. M., and Joh, T. H., 1978, Reversible changes in the activities and amounts of tyrosine hydroxylase in dopamine neurons of the substantia nigra in response to axonal injury as studied by immunochemical and immunocytochemical methods, *Brain Res.* **144**:325–342.

Rosner, B. S., 1974, Recovery of function and localization of function in historical perspective, in: *Plasticity and Recovery of Function in the Central Nervous System* (D. G. Stein *et al.*, eds.), Academic Press, New York, pp. 1–29.

Scheff, S. W., Bernardo, L. S., and Cotman, C. W., 1980, Decline in reactive fiber growth in the dentate gyrus of aged rats compared to young adult rats following entorhinal cortex removal, *Brain Res.* **199**:21–38.

Schoenfeld, R., and Uretsky, N., 1972, Altered response to apomorphine in 6-hydroxydopamine-treated rats, *Eur. J. Pharmacol.* **19**:115–118.

Schultz, W., and Ungerstedt, U., 1979, Spontaneous activity of striatum cells and their sensitivity to apomorphine in dopamine-lesioned rats, in: *Catecholamines: Basic and Clinical Frontiers*, Volume I (E. Usdin, I. J. Kopin, and J. Barchas, eds.), Pergamon Press, New York, pp. 631–633.

Seeman, P., 1980, Brain dopamine receptors, *Pharmacol. Rev.* **32**:229–313.

Severson, J. A., Osterburg, H. H., and Finch, C. E., 1981, Aging and haloperidol-induced dopamine turnover in the nigra-striatal pathway of C57Bl/6J mice, *Neurobiol. Aging* **2**:193–197.

Sharman, D. F., Poirier, L. J., Murphy, G. F., and Sourkes, T. L., 1967, Homovanillic acid and dihydroxyphenylacetic acid in the striatum of monkeys with brain lesions, *Can. J. Physiol. Pharmacol.* **45**:57–62.

Sokoloff, L., Reivich, M., Kennedy, C., Des Rosiers, M. H., Patlak, C. S., Pettigrew, K. D., Sakurada, O., and Shinohara, M., 1977, The [^{14}C]deoxyglucose method for the measurement of local cerebral glucose utilization: Theory, procedure, and normal values in the conscious and anesthetized albino rat, *J. Neurochem.* **28**:897–916.

Sokoloff, P., Martres, M. P., and Schwartz, J. C., 1980, Three classes of dopamine receptor (D-2, D-3, D-4) identified by binding studies with ^{3}H-apomorphine and ^{3}H-domperidone, *Naunyn-Schmiedeberg's Arch. Pharmacol.* **315**:89–102.

Sporn, J. R., Harden, T. K., Wolfe, B. B., and Molinoff, P. B., 1976, β-adrenergic receptor involvement in 6-hydroxydopamine-induced supersensitivity in rat cerebral cortex, *Science* **194**:624–626.

Staunton, D. A., Wolfe, B. B., Groves, P. M., and Molinoff, P. B., 1981, Dopamine receptor changes following destruction of the nigrostriatal pathway: Lack of a relationship to rotational behavior, *Brain Res.* **211**:315–327.

Stricker, E. M., and Zigmond, M. J., 1976, Recovery of function following damage to central catecholamine-containing neurons: A neurochemical model for the lateral hypothalamic syndrome, in: *Progress in Physiological Psychology and Psychobiology*, Volume 6 (J. M. Sprague and A. N. Epstein, eds.), Academic Press, New York, pp. 121–188.

Ungerstedt, U., 1971a, Stereotaxic mapping of the monoamine pathways in the rat brain, *Acta Physiol. Scand. (Suppl.)* **367**:1–48.

Ungerstedt, U., 1971b, Adipsia and aphagia after 6-hydroxydopamine induced degeneration of the nigro-striatal dopamine system, *Acta Physiol. Scand. (Suppl.)* **367**:95–122.

Ungerstedt, U., 1971c, Postsynaptic supersensitivity after 6-hydroxydopamine induced degeneration of the nigro-striatal dopamine system in the rat brain, *Acta Physiol. Scand. (Suppl.)* **367**:69–93.

Uretsky, N. J., Simmonds, M. A., and Iversen, L. L., 1971, Changes in the retention and metabolism of ^{3}H-1-norepinephrine in rat brain *in vivo* after 6-hydroxydopamine pretreatment, *J. Pharmacol. Exp. Ther.* **176**:489–496.

van der Krogt, J. A., and Belfroid, R. D. M., 1980, Characterization and localization of catecholamine-susceptible Na-K ATPase of rat striatum: Studies using catecholamine receptor (ant)agonists and lesion techniques, *Biochem. Pharmacol.* **29**:857–868.

Waddington, J. L., Cross, A. J., Longden, A., Owen, F., and Poulter, M., 1979, Apomorphine induced rotation in the unilateral 6-OHDA-lesioned rat. Relationship to changes in striatal adenylate cyclase activity and ^{3}H-spiperone binding, *Neuropharmacology* **18**:643–645.

Westlind, A., Grynfarb, M., Hedlund, B., Bartfai, T., and Fuxe, K., 1981, Muscarinic supersensitivity induced by septal lesion or chronic atropine treatment, *Brain Res.* **225**:131–141.

Yarbrough, G. G., 1976, Ouabain antagonism of noradrenaline inhibitions of cerebellar Purkinje cells and dopamine inhibitions of caudate neurones, *Neuropharmacology* **15**:335–338.

7

Do Rats Have Hypotheses?

A Developmental and Means–Ends Analyses Approach to Brain Damage, Recovery of Function, and Aging

M. LISA VALENTINO and DONALD G. STEIN

1. INTRODUCTION

Within the last few years, there has been an increasing interest in how organisms recover from central nervous system (CNS) injury, and there have been many reviews devoted to this question (Cotman, 1978; Finger, 1978a; Finger and Stein, 1982). Although the laboratory research has been exciting and there have been many new developments, clinicians have been slow to accept the idea that recovery after CNS injury is possible. Consequently, since little is done to aid the process, the clinical prognosis for recovery is usually rather poor. The prevailing pessimism concerning the victims of traumatic brain injury has been well summarized by Brailowsky (1980) in a recent volume on recovery from brain damage:

> Clinicians concerned with the physical therapy and general rehabilitation of persons with incapacitating brain lesions rarely make use of drugs to alter the course of the disease, and in general, the pharmacological management of these patients is only symptomatic. The majority of the ameliorative compounds originally came into use as a result of empirical observations, and with little or no understanding of the physiopathology involved. . . . In any case, the brain is considered so complex and

M. LISA VALENTINO • Department of Psychology, Clark University, Worcester, Massachusetts 01610.   DONALD G. STEIN • Department of Psychology, Clark University, Worcester, Massachusetts 01610, and Department of Neurology, University of Massachusetts Medical Center, Worcester, Massachusetts 01605.

delicate that any lesion sustained by this organ is usually and almost automatically given a grim prognosis. (page 187)

For the most part, recent investigators have been concerned with the question of whether regenerative or anomalous sprouting may be responsible for functional recovery after CNS lesions. Although there is not a consistent body of literature defining the behavioral concomitants of sprouting, in cases where some type of anomalous growth is seen, it is better in the young subject then in the old (Scheff *et al.*, 1978). At this point, we may ask if there might be any behavioral consequences of this new growth? Despite the possibility that neuronal sprouting in response to CNS injury mediates functional recovery, the specific behavioral patterns that permit the organism to adapt favorably to the demands of its environment have not been carefully defined (e.g., see Laurence and Stein, 1978, for discussion of the means–ends analysis problem).

Briefly stated, in order to study the question of how individuals respond to brain injury, it is necessary to evaluate whether the observed recovery is due to actual *sparing* of normal behavior (that is, is the postoperative behavior *identical* to the preoperative behavior?). Alternatively, one could ask whether the neuronal reorganization is manifested by a substitution of alternate behaviors or "tricks" (Goldberger, 1974) as a means of coping with the injury.

Although Sperry (1945) was one of the first investigators to concern himself with "substitution" as a behavioral mechanism for functional recovery, there really has been little systematic investigation on this issue since the now classic work of Lashley (1929) and Honzik (1936). These investigators were interested in determining which sensory system animals use most in solving a maze task. Honzik (1936) concluded that vision was the most important sense since maze learning was rapid as long as the visual system was intact. When he blinded his rats, they could still solve the maze, but they did so at a much slower rate. Through various analyses and manipulations, Honzik demonstrated that the rats that were unable to use visual cues depended on alternate cues such as olfaction. This switching from the dominant sensory system to a back-up system is what accounted for the apparent recovery in Honzik's brain-damaged rats. Both Honzik (1936) and Lashley (1929) reasoned that rats have a hierarchy of cues they will use and will fall back on less desirable strategies as their preferred ones are eliminated. Their work was confirmed at Clark University by Rosen and Stein (1969) who showed that adult rats will systematically select from a hierarchy of cues as they navigate through a T-maze.

About 50 years ago, Krechevsky (1932) also demonstrated that rats will use hypotheses, or strategies, to solve a discrimination task. He claimed that rats will not run in a "random or haphazard fashion" when they are required to solve a problem. Krechevsky argued that it is only the nature of the analysis of the behavior that makes it appear that the rats' performance is haphazard. He also pointed out that in traditional learning analyses, only the number of errors are usually recorded; therefore, unless the subject is responding according to the experimentally set criterion, or strategy, it would appear to the experimenter as if the animal responds randomly. Thus, if two subjects had error scores of 50%, it would seem as if they were responding randomly, but if the *patterns* of responding were examined, it might be observed that both subjects were responding in two entirely different but systematic ways. For example, this would be the case if in an alternation task, one subject responded LRRRRL and another responded RRLLLR.

In order to test his ideas, Krechevsky developed a task which enabled him to look at the patterns of responses that rats make when they are placed in a choice discrimination situation. For example, he was able to determine if the rats were responding to a left-side preference, right-side preference, dark preference, light preference, or any combination of these. Krechevsky tested a large number of rats in a four-unit alley maze and recorded their patterns of responses. He then looked for instances of responding to a cue, or stimulus, 70% or more of the time. He operationally defined this response rate as a hypothesis. The results supported his theory that rats, like humans, respond in a systematic way when faced with a problem.

With respect to brain-damaged subjects, Michael Goldberger (1972, 1974) has recently demonstrated that monkeys also use different hypotheses or "tricks" to recover from lesions of Brodmann's area 6 (supplementary motor area). Goldberger (1972) subjected monkeys to bilateral lesions of the supplemental motor area. After surgery the subjects displayed a pathological reflex of forced grasping to light tactile stimulation and an inability to release the grasped object. Goldberger attempted to teach his animals to hold and then release a stick for food reward, but he discovered that this pathological grasp could not be overcome in the period immediately following surgery. However, after a few weeks the monkeys showed improvement in overcoming the pathological grasp reflex, suggesting that some functional recovery had taken place. After additional tests, Goldberger rejected the idea of true functional restitution in favor of the hypothesis that the monkey's recovered behavior was actually due to a different behavioral strategy or trick (see Sperry, 1945, 1947).

In another recent study A. Gentile, S. Green, A. Nieburgs, W. Schmelzer, and D. Stein (1978) reported finding both substitution and restitution of function after brain injury, depending on the surgical manipulation employed. Serial or one-stage lesions of medial parietal cortex in the rat resulted in a pronounced impairment of locomotion on an elevated runway. Gentile *et al.* (1978) reported a difference in the patterns of hindlimb movements between recovered one- and two-stage subjects that was due to a substitution of alternate hindlimb movement patterns in rats with one-stage lesions, whereas the two-stage subjects' behavior appeared to be identical to the normal ballistic pattern of hindlimb placements.

Up to the present time, the question of what mediates recovery has been almost exclusively examined in very young or young adult organisms. Consequently, we are practically in the dark about the process of restitution of function (or whether such processes even exist!) in the aged subject, despite the fact that the aged are at a greater risk for degenerative and traumatic brain injuries. In one of the few papers examining behavioral recovery of function in aged subjects, Stein and Firl (1976) have suggested that senescent rats may be incapable of any recovery. These investigators argued that certain brain structures may change their functions or even become nonfunctional with increasing age. To test this notion Stein and Firl gave young adult and aged rats sham or one- or two-stage lesions of frontal cortex and tested them on spatial alternation and other tasks sensitive to frontal damage. The results that lead to the conclusion that brain structures can cease to mediate certain behaviors in aged subjects were based upon their observations that the aged subjects with lesions (one- and two-stage) did not differ from their aged matched controls on spatial alternation. However, when the normal aged rats were compared to normal young rats, it was apparent that the normal old rats were impaired in the amount of time it took them to learn an alternation scheme (see Fig. 1). Their histological analyses showed that there was also extensive cell loss in the frontal cortex and dorsomedial thalamic nucleus of the intact, aged rats. For example, the mean number of normal cells seen in the frontal cortex of the aged subjects was 20.0 as compared with a mean of 47.5 for the young subjects.

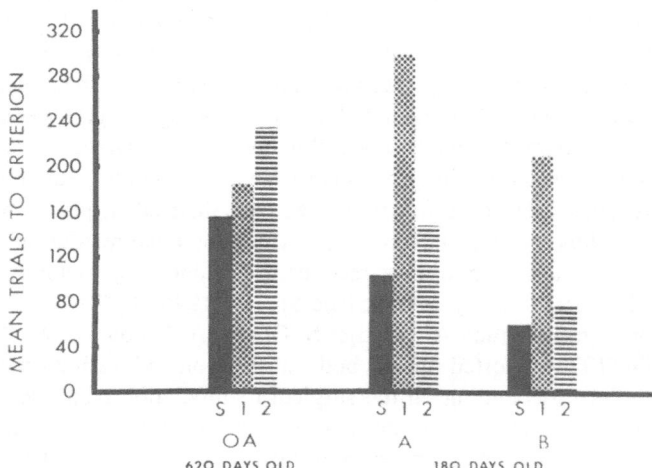

Figure 1. Delayed spatial alternation task (Stein and Firl, 1976). Spatial alternation performance in young adults or old rats with either lesions of frontal cortex or sham operations. The animals were tested to a criterion of 15/16 correct responses. Groups A and B were tested at different times and were therefore kept separate for the analyses.

Recently, Corwin, Vicedomini, Nonneman, and Valentino (1982), in a partial replication of Stein and Firl (1976), obtained different behavioral results. These investigators gave sham or one- or two-stage lesions of medial frontal cortex to young (35 days), adult (180 days), and aged rats (570 days) and tested them on a spatial alternation task. In this study the rats were given extensive interoperative testing and handling, which they believed would provide optimal

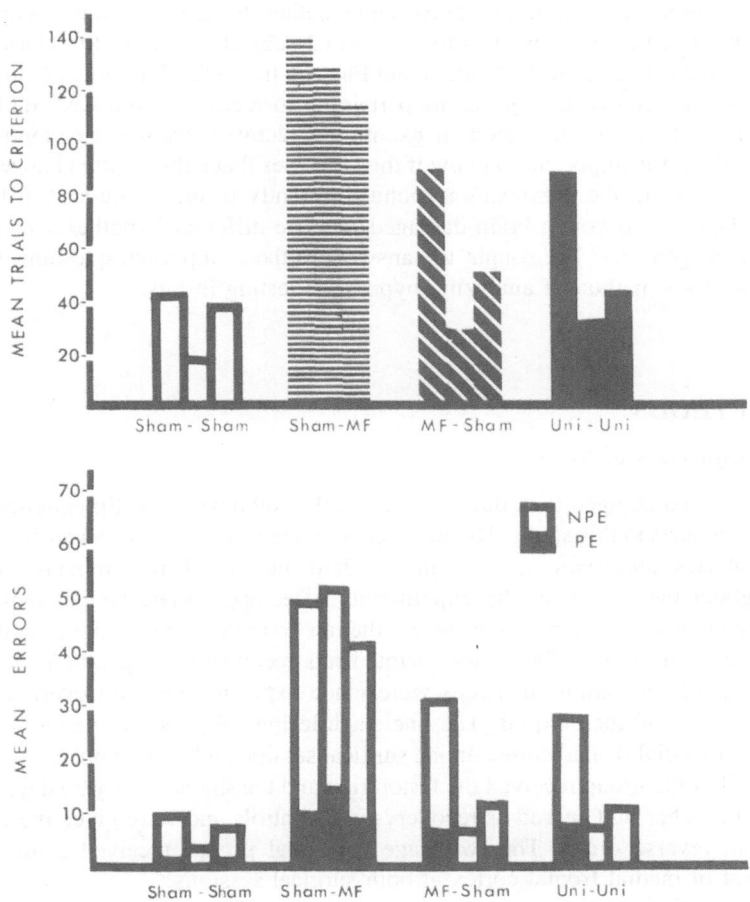

Figure 2. Mean number of trials to criterion (upper panel) and errors to criterion (lower panel) for the three age categories during retention testing on spatial-delayed alternation in the Corwin *et al.* (1982) study. The bars within each surgical category represent, from left to right, the 35-, 150-, and 570-day groups. Perseverative errors (PE) are indicated by the black portion of each error bar. Nonperseverative (NPE) or initial errors are indicated by the open portion of each bar.

conditions for recovery (see Finger, 1978a). All three age groups in the Corwin *et al.* paper demonstrated a serial lesion effect. The one-stage subjects were impaired, while the two-stage subjects were not, when compared to their age-matched sham controls. There also was no difference between the normal subjects across the age dimension (see Fig. 2).

The discrepancies in the results of these two studies led us to the present investigation. We were concerned with Lashley's (1929) and Krechevsky's (1932) original idea that animals, as well as people, use alternate "tricks" or strategies to solve learning problems, and whether the use of such tricks applies to the brain-damaged as well as to the normal subject. We hoped to clarify the discrepancy between the 1976 Stein and Firl and the 1982 Corwin *et al.* studies, and also answer two key questions pertaining to recovery from CNS damage.

First, if one is interested in examining recovery from a developmental perspective, it is important to know if the strategies that subjects use change with age. Second, do the organisms respond differently to brain injury at different ages? That is, do young brain-damaged rats use different hypotheses than old brain-damaged rats? We sought the answers to these important questions using Krechevsky's method of analyzing hypothesis testing in rats.

2. METHODS

2.1. Subjects and Surgery

We used 20 aged (570 days) and 21 adult (180 days) male Sprague-Dawley rats as subjects in this study. The rats were obtained from Charles River Breeding Laboratories and were housed in standard individual rack-mounted cages throughout the course of the experiment. After appropriate weight reduction required to get them to run in the maze, the rats were randomly assigned to one of three surgical groups. The sham-operated rats received two operations 30 days apart, in which midline incisions were made exposing the skull overlying the frontal pole and then closed. The one-stage lesion subjects received a bilateral lesion of medial frontal cortex at one surgical session and a sham operation at the other. Half the group received the lesion first and the sham surgery 30 days later, while the other half served as recovery-time controls and were given the operations in reverse order. The two-stage, or serial group, received a unilateral removal of medial frontal cortex at both surgical sessions.

The lesions of medial frontal cortex were done by gentle aspiration so as to remove the midline tissue from bregma to the frontal pole, until the dorsal surface of the olfactory bulb and tract could be visualized anteriorly and the corpus callosum posteriorly.

2.2. Behavioral Testing

Prior to the actual experiment, the rats were allowed 30 days to adjust to the laboratory environment. During this time they were maintained on *ad libitum* food and water. The rats were tested in the Krech box for two 14-day sessions, one occurring prior to any surgery, the other postoperatively. Both testing sessions were identical.

All of the rats were given two weeks of pretraining prior to the first testing session and then seven days prior to the second testing session. Pretraining consisted of placing a rat in the goal box, with reward present, for five minutes a day.

Since the Krech box is not used frequently, a brief description may prove helpful to the reader. The "Krech" (1932) box is a long straight alley consisting of a start box, four discrimination units, and a goal box. The discrimination units consist of two alleys, one of which was blocked during each trial. Each alley is equipped with a light which could be on or off; only one alley was lit within each unit during a trial. The position of the blocked alley and lighted alley varied within each unit, and over each trial, according to a random schedule. The rats are required to trasverse all four units during each trial. The animals were rewarded at the end of every trial.

The rats were given 12 trials per day for 14 days.* The position of the alley which the subject first entered was recorded by the experimenter for each unit, regardless of whether it was correct or incorrect. If the animal entered the incorrect alley, it was required to retrace and enter the correct alley in order to enter the next unit. Thus four responses were recorded per trial, or 48 per day. Once the subjects exited a unit, they were not allowed to reenter that unit until the next trial.

The subjects' responses were compared with schedules of responses for particular cues; for example, left responding, dark responding, or an alternation scheme. (Details concerning these schedules are given later in this chapter.) The data were analyzed for eight such cues. The subjects were said to be using a particular cue, or attending to a particular strategy, if 70% of their responses were the same as the cue schedule. The strategy analyses were done using an Apple II computer.

All of the rats were given their first surgery (sham or aspiration of medial frontal cortex) the day following the completion of the preoperative testing session as described above. They were then permitted a 30-day interoperative recovery interval which was followed by a second surgery (sham or tissue re-

---

*The rats were actually tested for 15 days per session, since the 12 trials of day 1 in each testing session were given in two six-trial days.

moval surgery). After the second surgery the rats were given a 10-day rest period, before the start of the second testing session. The postoperative testing session was identical to the preoperative session.

2.3. Histology

At the completion of all testing the rats were killed and prepared for histology. Their brains were then embedded in egg yolk and cut into frozen sections 30 μm thick. Every sixth section was mounted and stained with cresyl-violet acetate, and lesion extent was determined for all subjects. (See Figs. 6 and 7 for the maximum/minimum lesion in each surgical group.)

2.4. Data Analysis

As mentioned previously, the individual responses of the rats were compared with eight schedules of responses to a particular cue. The cue schedules are as follows: Left responding, right responding, light responding, dark responding, spatial alternation, spatial perseveration, visual perseveration, and the schedule of correct responses.

The responses required by the particular cue schedule should be reported here. In order to respond according to a "right" hypothesis, the subject must respond to the right alley position, whereas a "left" hypothesis would necessitate responding to the left alley. Responding to the dark alley regardless of spatial location would be in keeping with a dark hypothesis. A light hypothesis would then mean solely responding to the light cue. The spatial alternation schedule was such that the rats need to respond to the spatial opposite of the correct alley in the previous unit. For example, if left was correct in unit 1, then to keep with an alternating scheme, the subjects would need to respond right in unit 2; they would then respond to the opposite of what was correct in unit 2, in unit 3, etc. The spatial perseveration scheme required the rat to respond to the correct spatial cue in the previous unit; therefore, if right was correct in unit 2 and the rat responded right in unit 3, this act would be considered in keeping with a spatial perseveration hypothesis. The visual perseveration hypothesis used the same principle with visual cues. Thus, if dark was correct in unit 1, the rat would need to respond to dark in unit 2, and so on.

The scores generated in the analysis represented the number of responses, out of the 48 total responses, that were made to the particular cue.* An arbitrary

---

*It should be pointed out that when the rats were using an alternating or perseverative hypothesis, their responses in unit 1 were irrelevant since there was no previously correct response. Thus, there were only 36 total possible responses for these hypotheses on each day. This was taken into account when the percentages were computed.

criterion of 70% responding to a cue was defined as a hypothesis, or strategy. This criterion was chosen because it was used by Krechevsky,* and was considered to reflect or demonstrate an unambiguous use of an hypothesis by the animal.

3. RESULTS

We felt that it was important to look at the individual subject's response to the brain injury. We found that the most efficient way in which to do this was to present the behavioral patterns of each subject in graphic form (see Figs. 3, 4, and 5 for representative subjects). The graphs represent the percentage of responding on each day to all the cues in which the criterion of 70% was met on any day. The 15 days of preoperative and 15 days of postoperative performance are separated by an open space in the graph.

If one carefully examines the individual performance profile, it seems apparent that the normal-aged rat is much more variable in its strategy testing behavior than younger counterparts (see Fig. 3). This can be seen by the number of sharp peaks in the percentage of responding to the cues available in the long runway. Conversely, the data of the normal young subjects show more extended "plateaus" where the rats appear to be perseverating on a particular cue for an extended period of time, rather than shifting from one strategy to another as the older rats do. If the mean number of days per hypothesis is counted, we can see the same phenomenon, namely that the younger rats are more focused in their strategy-testing behavior than their aged counterparts (see Table 1).

While the aged subjects spent less time per hypothesis than the younger rats, there was no difference in the total number of hypotheses tested between the two age groups. Also, although neither age cohort had an overall preference for spatial over visual cues, or vice versa, some of the individual rats did respond to spatial or visual cues preferentially.

Postoperatively, as expected, the normal (sham) young subjects continue to perseverate for long periods of time on a particular strategy, as they did prior to the surgery. In contrast, the aged shams begin to show the long periods of perseveration only after an extended period of exposure to the test situation (see Fig. 3 and Table 1).

With respect to surgery, one-stage removals of medial frontal cortex resulted in the subjects in both age groups being more variable in their strategy-testing behavior. Thus, old rats and young rats with one-stage, bilateral surgery spent less days per hypothesis and showed shorter periods of testing the same

*Ivan Krechevsky changed his name to David Krech sometime during the 1930s.

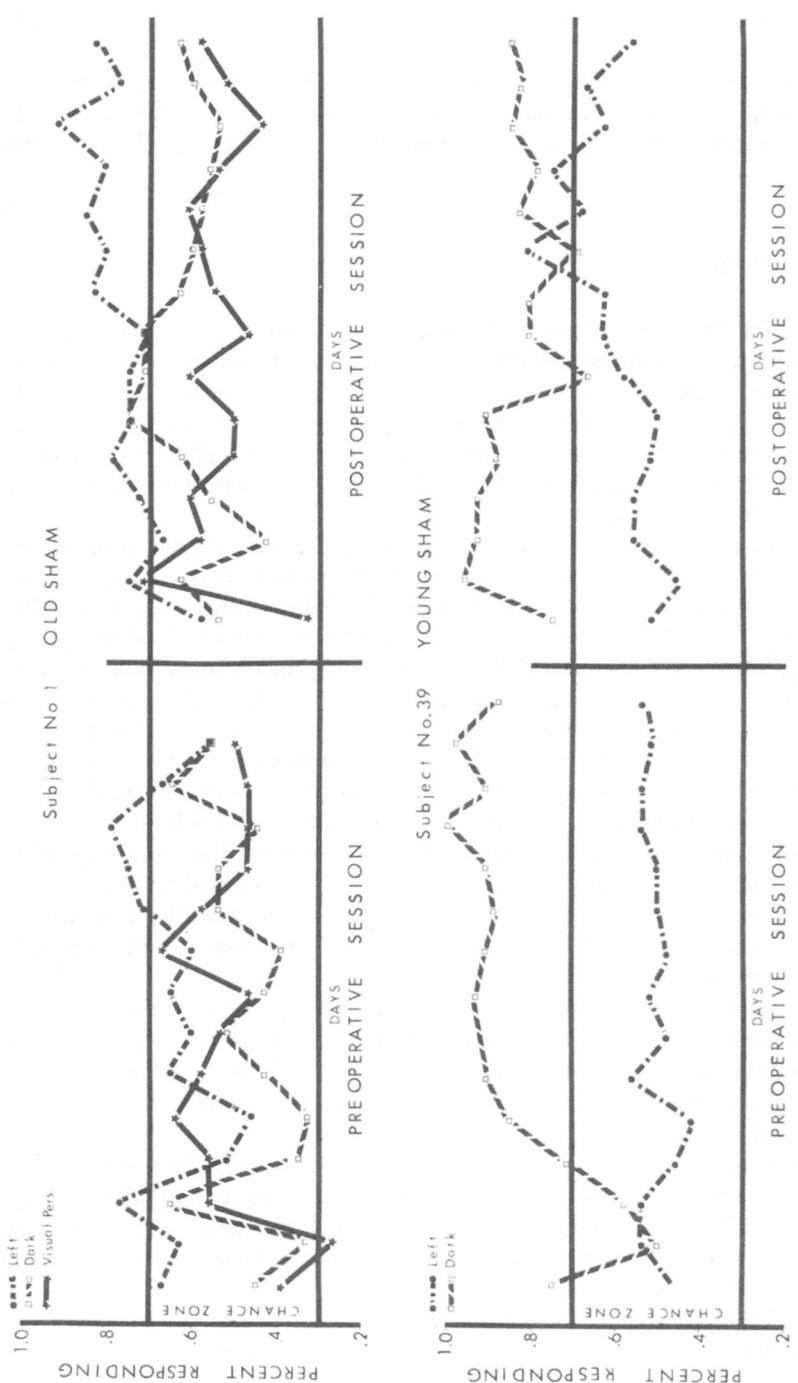

Figure 3. Pre- and postoperative strategy behavior of representative young and old sham subjects. Data points represent the percentage of responding to the particular cue for each day of the testing session. Points above 70% or below 30% represent a "hypothesis."

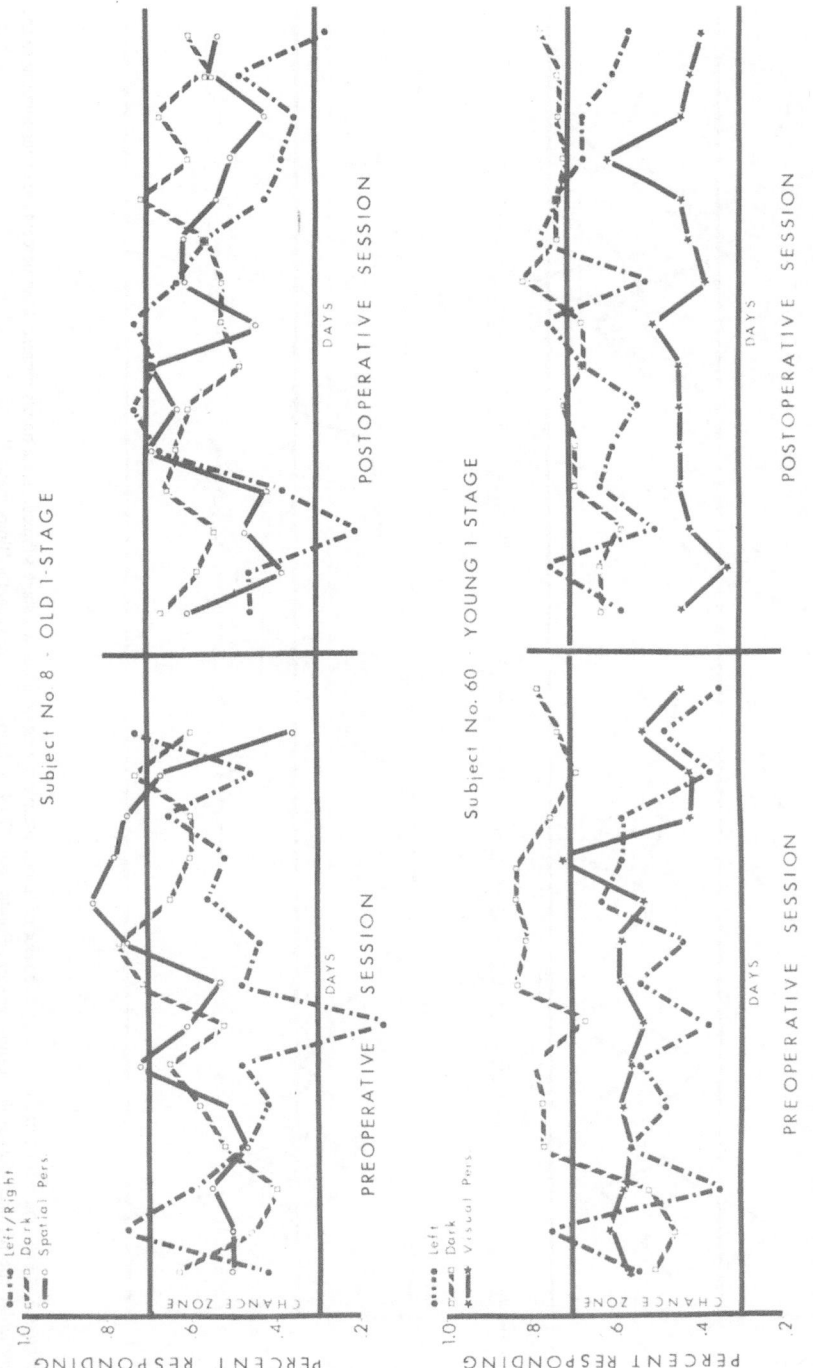

Figure 4. Pre- and postoperative strategy behavior of representative young and old one-stage subjects. Data points represent the percentage of responding to the particular cue for each day of the testing session. Points above 70% or below 30% represent a "hypothesis."

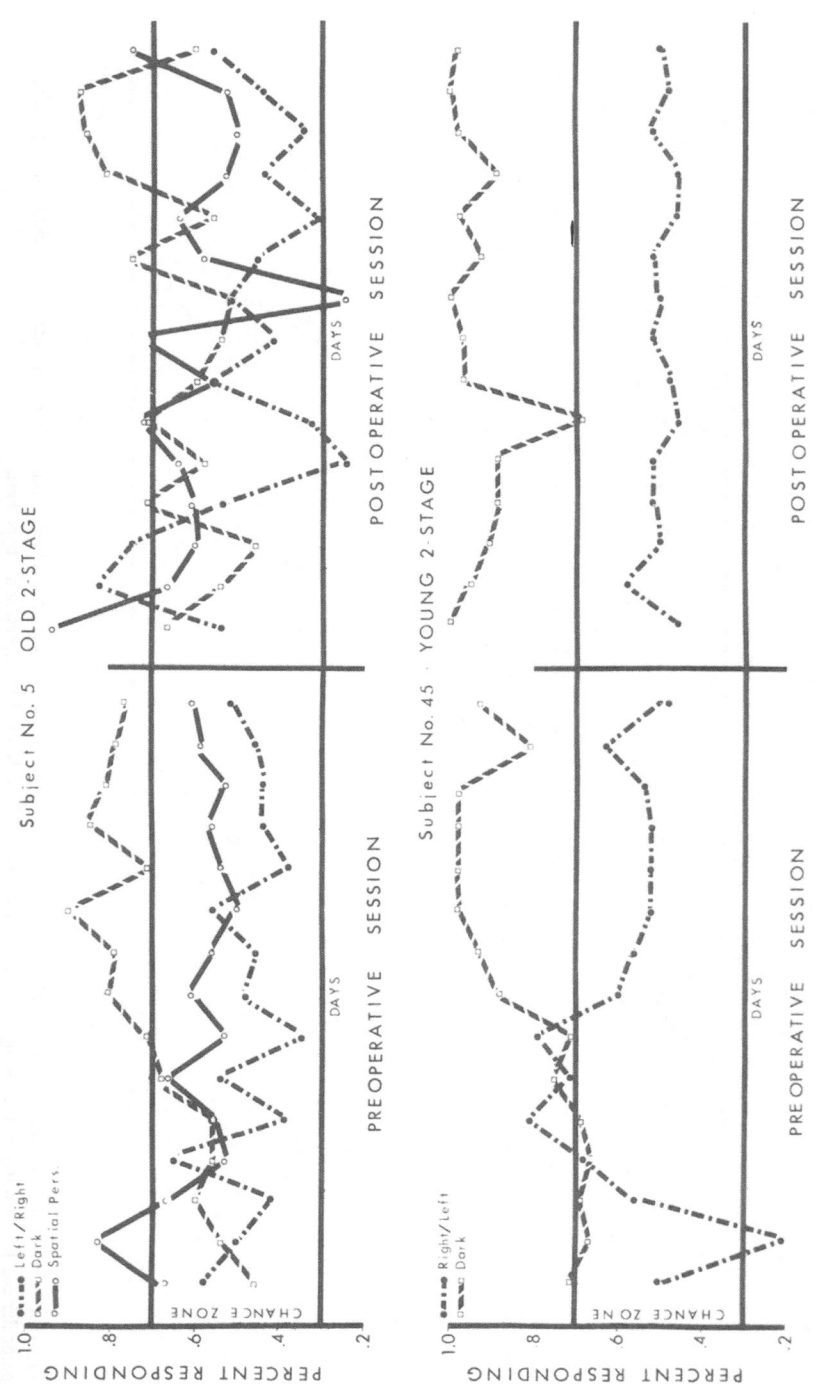

Figure 5. Pre- and postoperative strategy behavior of representative young and old two-stage subjects. Data points represent the percentage of responding to the particular cue for each day of the testing session. Points above 70% or below 30% represent a "hypothesis."

Table 1. Mean Number of Days Spent per Hypothesis

Subject group	Young	Old
Preoperative		
Normal	6.82	4.00
Postoperative		
Sham	11.46	7.00
1-Stage (40-day recovery time)	5.60	2.32
1-Stage (10-day recovery time)	5.60	5.60
2-Stage	10.00	4.00

hypothesis (i.e., more peaks) than their sham-operated counterparts ($\bar{X}_{\text{days old}}$ = 2.32; $\bar{X}_{\text{days young}}$ = 5.6; see Fig. 4).

In the younger subjects no difference was observed in the groups allowed different recovery times ($\bar{X}_{\text{days young, 40-day recovery time}}$ = 5.6; $\bar{X}_{\text{days young, 10-day recovery time}}$ = 5.4). This was not the case in the older subjects; it appears that the 40-day recovery time group was more variable (i.e., fewer days per hypothesis, more peaks) than their 10-day counterparts ($\bar{X}_{\text{days old, 40-day recovery time}}$ = 2.32; $\bar{X}_{\text{days old, 10-day recovery time}}$ = 5.6). It is felt by the experimenters that this result may be due to a small number ($N = 2$) in the 10-day-recovery-time group. This problem occurred due to extensive loss of subjects in this group.

Two-stage removals of medial frontal cortex in the aged rats resulted in the subjects remaining highly variable in their strategy behavior, although they were not as variable as the one-stage lesion subjects. In keeping with findings showing that two-stage lesions do not produce the same symptoms as damage inflicted in a single operation, two-stage lesions did not affect the strategy behavior of the younger animals. These subjects continued to show long periods of testing a particular hypotheses ($\bar{X}_{\text{days per hypothesis}}$ = 5.6; see also Fig. 5).

3.1. Qualitative Results

Throughout the course of this investigation certain qualitative differences were noted in the behavior of the rats in the different age groups. It is felt that these differences may have an important bearing on the differences in strategy-testing behavior across the age continuum, and that they should be reported here.

For example, we noted that the aged animals were more lethargic in the maze situation, taking much longer to negotiate the runway. The old rats required coaxing to retrace and enter the correct alley when they entered a blocked alley. In contrast, the younger animals traversed the maze very rapidly, rarely needing any coaxing. The aged rats were more deliberate in their choices, taking extended amounts of time to investigate all cues at the choice points, while the younger subjects appeared to have "tunnel vision," attending to a single cue and seemingly unaware of any others.

3.2. Medial Frontal Lesion Analysis

All brain-damaged subjects in both age groups sustained complete lesions of the medial frontal zone as defined by Leonard (1969). There were no significant differences in lesion size between any of the lesion or age groups. While most of the rats received some damage to the genu of the corpus callosum, in no subject was it totally transected. Twenty-five percent of the aged and 33.3% of the young subjects sustained some olfactory bulb damage. No subjects sustained damage to the head of the caudate nucleus. Figures 6 and 7 illustrate the maximum and minimum sizes of the medial frontal lesions for each lesion group of old and young rats.

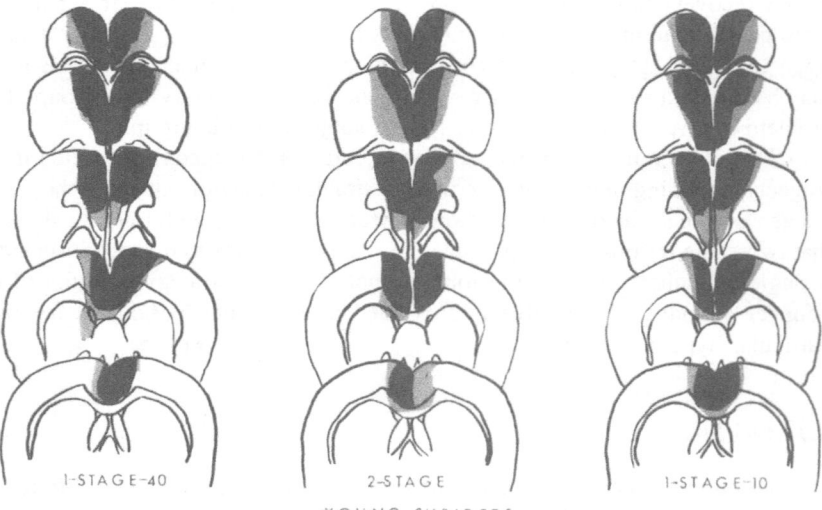

Figure 6. Maximum and minimum size of medial frontal lesions in the young subjects for each surgical group. The minimum lesion is indicated by the blackened areas and the maximum lesion is equivalent to the blackened areas plus the lined areas. 1-Stage-40, 1-stage MF lesion given 40 days recovery time; 2-stage, 2-stage MF lesion; 1-stage-10, 1-stage MF lesion given 10 days recovery time.

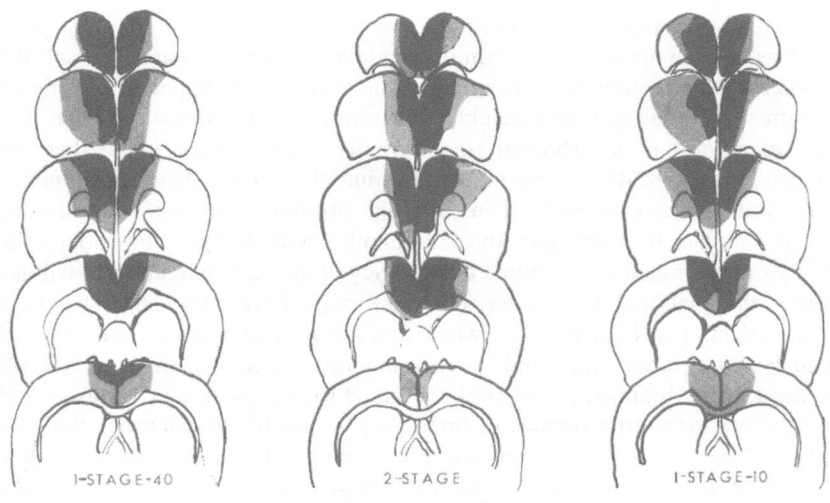

Figure 7. Maximum and minimum size of medial frontal lesions in the old subjects for each surgical group. The minimum lesion is indicated by the blackened areas and the maximum lesion is equivalent to the blackened areas plus the lined areas. 1-Stage-40, 1-stage MF lesion given 40 days recovery time; 2-stage, 2-stage MF lesion; 1-stage-10, 1-stage MF lesion given 10 days recovery time.

4. DISCUSSION

Our results can be taken to indicate that normal, aged rats are more variable in their hypothesis-testing behavior than normal young rats. We also noted that the aged subjects took longer to negotiate the alley maze and were much more deliberate in their choice behavior (i.e., they appeared to examine the cues at each choice more thoroughly than younger counterparts).

Although the normal, aged subjects improved in the efficiency with which they navigated the runway, aged brain-damaged rats—regardless of whether their medial frontal cortex lesions were made in one or two stages—remained highly variable. These rats never developed a consistent approach to running in the alley that characterized the performance of their younger counterparts.

Young adult rats with one-stage medial frontal lesions, like older "normal" rats, tended to switch from one hypothesis to another in rapid succession, whereas those with serial lesions developed strategies early in the testing and used them consistently to traverse the Krech box more efficiently and in a manner basically indistinguishable from intact controls.

Our observations have confirmed Krechevsky's (1932) notion that young intact rats do not approach a discrimination in a haphazard fashion; instead, they respond in a systematic way, testing various available strategies or hypotheses, one after the other and then quickly selecting the single hypothesis that leads them to reward in the shortest time. We have also demonstrated that aged subjects, too, have this ability to attend to individual cues, but they do not use these strategies as efficiently. Thus, the old rats and young rats with one-stage lesions vacillate from one hypothesis to another without spending enough time on any one approach to determine the saliency of the particular cue that will lead them to the goal box. The increased variability (or lower efficiency) seen in the aged subjects can explain why, when traditional error analyses are used, old animals appear to be "impaired" on learning tasks. It may be the case that since the aged or the brain-damaged subjects attend to particular cues (or use certain hypotheses) for shorter periods of time, they are unable to determine the "correctness" or "incorrectness" of any given cue as effectively. Therefore, they may test an incorrect cue on a number of separate occasions, never determining its importance for the task at hand.

Our observations lead us to suggest that young intact rats will attend to a cue (or specific set of cues) long enough to determine if it is "appropriate." If the cue is determined to be inappropriate, the rat will discard the hypothesis for another one (see the young subjects' data in Figs. 3, 4, and 5 for an example of consistent cue selection and efficient utilization). Such consistent behavior allows the young intact subject to find and use the correct response strategy in a shorter period of time. This would be especially evident if the required response was high on the hierarchy of cues that Lashley (1929), Honzik (1936), and Rosen and Stein (1969) have previously discussed.

One possible explanation for increased variability seen in the aged rats is that due to a loss in the number of sensory and motor neurons in the spinal cord (Mufson and Stein, 1980; Campbell et al., 1980), less information about the environment is being transmitted and processed by the CNS. Along with this decrease in spinal neurons, investigators have also reported significant losses in the number of neurons in cortical and subcortical structures throughout the brain (Mufson and Stein, 1980; Sabel and Stein, 1981; Hall et al., 1975; Bondareff, 1981). In one recent paper Elliot Mufson and Donald Stein, like Stein and Firl (1976) (see present introduction), reported an age-related reduction in the number of neurons in the frontal cortex of the Fischer 344 rat. They also observed other morphological changes as rats became senescent including: distortion of the neural soma, lateral displacement of the nucleolus, and an accumulation of lipofuscin material, events all associated with aging and cell death.

Cortical and spinal neurons are not the only cells vulnerable to age and injury-induced loss. For example, Sabel and Stein (1981) recently examined the number and size of neurons in several subcortical structures (e.g., septum, ventromedial and lateral hypothalamus, cortical and lateral nuclei of the amygdala) in young (90 days) and senescent (880 days) rats. These authors found a 10% brain shrinkage and extensive neuronal loss in all but the amygdala (see Table 2).

The findings from these anatomical studies indicate that along with the decreased number of neurons in the spinal cord, the feedback loop of association fibers within the brain may also be disrupted. This leads us to the conclusion that behavioral variability in the aged rats may be due at least in part to their inability to become sufficiently activated by, and to recognize and integrate, sensory stimuli.

In his theory of the role of experience and how experience functions in learning, Fiske and Maddi (1961) argued that there is an optimal level of activation necessary for any effective behavioral performance. Fiske defines activation as the state of the organism along the dimension of attention and arousal. According to him, activation is affected by the variation, intensity, and meaningfulness of a stimulus. Fiske argues that until the stimulation is intense enough for the threshold of activation to be reached, the organism may be "inattentive, easily distracted and not concentrating fully on the task" (page 32). He also points out that the animal will modify its level of activation by exploration of the environment until the optimal level is reached.

Thus, if we consider both Fiske's theory and the research on cell loss in the senescent rat, it is not surprising that the aged subjects in our study were more variable than their younger counterparts. Given the extensive neuronal loss seen in the aging brain, we can speculate that older rats are not receiving the same amount of sensory information as younger animals, and therefore their activation level is not optimum for efficient problem solving. In order to reach "activation threshold" with less efficient sensory and motor "apparatus," the aged subject must increase its exploration of the environment. More exploration is then seen as increased variability in responding. This variability is usually "factored out" by careful selection of statistical tests.

Many investigators have also reported dramatic morphological changes in the aged human CNS. For example, Scheibel and Scheibel (1975) have argued that pathological changes in the human prefrontal cortex are correlated with cognitive, mnemonic, and motor dysfunctions, as well as senile dementia.

Degenerative changes have also been demonstrated in the motor (Scheibel, 1980) and sensory (Brody, 1970) systems in humans. Scheibel (1980) reports that among the most obvious changes in the motor neurons of the spinal cord are swelling of the cell body, eccentricity of the nucleus, and engorgement of the cell body with lipofuscin granules. He reports that in a sample of aged patients, 60–90% of the neurons contain lipofuscin in sufficient quantities to compress the nucleus against the cell wall. He also reports a decrease in the number of small (< 15 μm) and medium-size (15–40 μm) neurons.

Table 2. Mean Total Number of Neurons for the Different Age Groups in Sabel and Stein (1981)[a]

Age (days)	VMH (%)		SL (%)		SN (%)		ACO (%)		ALP (%)	
90	58,635	(100.0)	616,684	(100.0)	224,383	(100.0)	213,388	(100.0)	119,404	(100.0)
360	57,638	(98.3)	460,492	(74.7)	155,876	(69.5)	166,404	(78.0)	139,512	(116.8)
520–580	50,601	(86.3)	332,002	(53.8)	190,930	(85.1)	160,540	(75.2)	135,513	(113.5)
850–880	23,086	(39.4)	197,579	(32.0)	100,992	(45.0)	112,596	(52.8)	92,521	(77.5)
F	7.03		4.57		6.28		5.14		2.57	
P	<0.004		<0.02		<0.007		<0.01		ns	

[a]VMH, ventromedial hypothalamus; SL, dorsal lateral septum; SN, substantia nigra; ACO, cortical amygdala; ALP, lateral amygdala. ns, Not significant. The percentages of remaining neurons compared with the youngest group are in parentheses. F values and levels of significance calculated by a one-way analysis of variance (ANOVA).

Taking the morphological changes together with Fiske's theory, we can speculate that the reason some older patients appear confused or disoriented may be due to the decreased number of sensory and motor neurons and the pathological changes in the cortex. Such patients cannot reach their appropriate "activation threshold." Since these patients are not receiving and/or integrating the external information, they cannot respond appropriately, and thus may appear disoriented. These individuals may also attempt to increase their external stimulation by increasing the variability of their actions, causing them to appear confused. This notion has been further tested in human patients by Klisz and Parsons (1977). These investigators examined hypothesis testing in young and old alcoholics on a task designed by Levine (1966). Their results indicated that older alcoholics use less systematic approaches to problem solving (i.e., they are significantly more variable than normals). Courville (1955) suggested that alcoholism may cause premature aging of the brain. Although the exact locus of this deterioration still causes some debate, Oscar-Berman (personal communication) has speculated the frontal lobe damage may be involved in producing the same kind of cognitive deficits as those seen in Korsakoff patients.

In the present study, Fiske's theory on the relevance of attention and experience may also explain the change in the behavior of the younger subjects after they received one-stage bilateral lesions of the medial frontal cortex. It is possible that by removing this cortical tissue, the integration loop which analyzes the external environmental cue has been disrupted or destroyed (Gentile et al., 1978).

Therefore, although the young subject with brain damage receives the environmental information at the receptor level, the information cannot be integrated. Thus, the brain-damaged subject is unable to respond efficiently.

Our data also indicate that serial lesions in young subjects spare the animals' strategy-testing behavior. It has been known since the early 1800s that slow-growing lesions often have less deleterious effects on the behavior of patients than damage of rapid onset. This idea was referred to as the "momentum of lesion" effect by Jackson (1879) and Finger (1978a).

To date, behavioral sparing after CNS injury using the sequential lesion paradigm has been demonstrated by many investigators (Ades and Raab, 1949; Fass et al., 1975; Nonneman and Kolb, 1979; Corwin et al., 1981, 1982).

Despite these numerous demonstrations, no good explanation for this functional sparing has been found. However, it might be suggested that external environmental stimulation during the interoperative interval permits the animal to "cope" better. It has been shown in a number of experiments that stimulation during the interoperative interval facilitates recovery from brain lesions. For example, Meyer, Isaac, and Maher (1958) demonstrated that a certain level of sensory stimulation is needed if rats with serial lesions of visual cortex are to be spared in a preoperatively learned-avoidance habit. Handling and specific training during the interoperative interval will also enhance performance after serial lesions (see Finger *et al.*, 1973; Stein, 1974; Finger, 1978b; and Finger and Stein, 1982, for reviews).

It is possible then that stimulation during the interoperative interval may serve to increase activation level, or arousal, of the subject. This increase in the level of arousal (or environmental enrichment) would then allow the rats to perform better in the test situation.

Research on amphetamine-facilitated recovery from brain damage supports the notion that increased activation enhances recovery from brain damage. For example, Braun, Meyer, and Meyer (1966) reported that amphetamine facilitates the reacquisition of a brightness discrimination in a Lashley maze. Subsequently, Meyer, Meyer, and Ritchie (cited in Meyer, 1974) noted that cats with total visual neocortex lesions given amphetamine injections were able to place their front paws accurately on the edge of a table when they were moved slowly toward the edge. When the effect of the amphetamine wore off, the visual defects reappeared. The cat's ability to discriminate edges and borders lasted as long as the drug was maintained. In another experiment, Meyer, Horel, and Meyer (1963) reported that the immediately noticeable effects of amphetamine injections were an increase in the rate of volume of respiration, piloerection of the fur on the animals' back and tail, and other manifestations of high arousal. Thus, we can speculate that amphetamines may enhance the brain-damaged subject's ability to reach "activation threshold," in Fiske's terms.

This idea receives further support from a study by Simone Faugier-Grimaud, Colette Frenois, and Donald Stein (1978). These investigators examined visually guided placing reactions after removal of posterior parietal cortex lesions in adult Java monkeys. After surgery, the animals demonstrated a deficit similar to Balint's syndrome, but recovered from this deficit in ten days to two weeks. Approximately one year after surgery, Faugier-Grimaud *et al.* gave the monkeys a small dose (5% of the original dose) of the anesthetic used for the surgery (ketamine). Under the drug condition the investigators noted an immediate return of the original deficit and as the drug wore off, the impairment of eye-hand-mouth coordination disappeared. It appeared as if the anesthesia prevented the monkeys from obtaining the required optimum level of arousal. As Faugier-Grimaud *et al.* stated in their conclusions:

> We should point out that almost 30 yr ago, Lashley (1950) speculated that "amnesia from brain injury rarely, if ever, is due to destruction of specific memory traces. Rather, the amnesias represent a lowered level of vigilance, a greater difficulty in activating the organized patterns of traces, or a disturbance of some broader system of organized functions. (p. 166)

Returning to our present study, in light of the argument that a certain level of activation is necessary for performance, it is not surprising that there was no difference in the efficiency with which the aged, brain-damaged subjects approached the strategy testing task, regardless of whether the lesions were produced in one or two stages. Since the normal aged subjects may already require a higher activation threshold due to cellular and neuronal loss, serial damage to the association cortex does not confer any benefit. Actually, the additional neuronal loss from the lesion, regardless of speed of onset, makes the aged rat even more unable to respond to the cues available in the maze situation. These results, along with the anatomical studies on the aged brain, led us to the conclusion that Stein and Firl (1976) were correct when they stated that ". . . the functions of the frontal area change in some as yet unspecified manner so that this region no longer contributes to the mediation of spatial behaviors" (page 165).

As pointed out in the introduction, Corwin (1982) and his colleagues at the University of Kentucky demonstrated a serial lesion effect in their aged rats with frontal lesions. Even though these results differ from those of Stein and Firl (1976) and the present study, in light of the argument presented here, they can be expected. As noted previously, Corwin *et al.* subjected both young, adult, and senescent rats to one- or two-stage lesions of the medial frontal cortex. In contrast to Stein and Firl, the experimenters gave their subjects extensive interoperative training and handling. Corwin *et al.* argued that these conditions were optimal for recovery, and this idea receives support from Finger and Stein (1982) and from others who have examined the beneficial role of enriched or complex environments on recovery from brain damage.

This increased experience and varied stimulation, e.g., handling and specific training, may have served to allow the older rats to attain and maintain their activation threshold without exploring the maze environment repeatedly, and in great detail. Therefore, the aged rats who had been given extensive handling by Corwin *et al.* were less easily distracted and thus could attend to the salient cues in the learning task more readily. With the advantage of having attained activation threshold, these subjects need not explore their environment extensively at the start. We could then expect the older rats to be able to compensate for their injury and thus demonstrate a serial lesion effect, as was the case in the Corwin *et al.* study.

In conclusion, we feel that it is important to use an individual "means–ends analysis" (Laurence and Stein, 1978), such as the one employed here, if we are to understand how an individual responds to brain injury. Traditional analyses using group means, due to an underlying assumption that any within-group deviation from the mean is due to chance or error in measurement, may, and often does, mask any subtle differences in the way individuals respond to their environment and to brain injury. The Krech box allows us to examine these subtle differences by observing the individual response strategies of the subject. It seems especially important to know how an organism uniquely responds to brain injury (i.e., what are the coping strategies used?) in developing rehabilitation and treatment tailored to the individual case. Hopefully, armed with more sophisticated "means–ends analyses" of behavior and brain function, we may someday develop the techniques necessary to change the pessimistic view that permanent disability is all that is available to the brain-damaged individual.

ACKNOWLEDGMENTS.   We would like to express sincere thanks to Ms. Judith Cassimeris for her assistance in the behavioral testing. Gratitude is also extended to Mr. Arthur C. Firl for his help in preparation of the manuscript and to Mrs. Mildred Sanders for the typing of the manuscript. This research was supported in part by the National Institute of Aging (Grant #AG00295), the United States Army Medical Research and Development Command (Contract #DAMD17-82-C-2205), and the Department of Psychology at Clark University. This research was done in partial fulfillment of the requirements for the master's degree.

REFERENCES

Ades, H. W. and Raab, D. R., 1949, Effects of preoccipital and temporal neodecortication on learned visual discrimination in monkeys, *J. Neurophysiol.* **12**:101–108.

Bondareff, W., 1981, The neurobiological basis of age-related changes in neuronal connectivity, in: *Aging: Biology and Behavior* (J. L. McGaugh and S. B. Kiesler, eds.), Academic Press, New York, pp. 141–158.

Brailowsky, S., 1980, Neuropharmacological aspects of brain plasticity, in: *Recovery of Function: Theoretical Considerations for Brain Injury Rehabilitation* (P. Bach-y-Rita, ed.), University Park Press, Baltimore, pp. 187–224.

Braun, J. J., Meyer, P. M., and Meyer, D. R., 1966, Sparing of a brightness habit in rats following visual decortication, *J. Comp. Physiol. Psychol.* **61**:79–82.

Brody, H., 1970, Structural changes in the aging nervous system, *Interdiscip. Top. Gerontol.* **7**:9–21.

Campbell, B. A., Krauter, E. E., and Wallace, J. E., 1980, Animal models of aging: Sensory-motor and cognitive function in the aged rat, in: *The Psychobiology of Aging: Problems and Perspectives* (D. G. Stein, ed.), Elsevier/North-Holland, New York, pp. 201–226.

Corwin, J. V., Nonneman, A. J., and Goodlet, C., 1981, Limited sparing of function on spatial delayed alternation after two stage frontal cortex lesions in the rat, *Physiol. Behav.* **26**:763–771.

Corwin, J. V., Vicedomini, J. P., Nonneman, A. J., and Valentino, L., 1982, Serial lesion effect in rat medial frontal cortex as a function of age, *Neurobiol. Aging* **3**:69–76.

Cotman, C. W. (ed.), 1978, *Neuronal Plasticity*, Raven Press, New York.

Courville, C., 1955, *Effects of Alcohol on the Nervous System of Man*, San Lucas, Los Angeles.

Fass, B., Jordan, H., Rubman, A., Seibel, S., and Stein, D. G., 1975, Recovery of function after serial or one-stage lesions of the lateral hypothalamic area in rats in *Behav. Biol.* **14**:283–294.

Faugier-Grimaud, S., Frenois, C., and Stein, D. G., 1978, Effects of posterior parietal lesions on visually guided behavior in monkeys, *Neuropsychologia* **16**:151–168.

Finger, S., 1978a, Lesion momentum and behavior, in: *Recovery from Brain Damage: Research and Theory* (S. Finger, ed.), Plenum Press, New York, pp. 135–164.

Finger, S., 1978b, Environmental attenuations of brain lesion symptoms, in: *Recovery from Brain Damage: Research and Theory* (S. Finger, ed.), Plenum Press, New York, pp. 297–329.

Finger, S., and Stein, D. G., 1982, *Brain Damage and Recovery: Research and Clinical Perspectives*, Academic Press, New York.

Finger, S., Walbran, B., and Stein, D. G., 1973, Brain damage and behavioral recovery: Serial lesion phenomena, *Brain Res.* **63**:1–18.

Fiske, D. W., and Maddi, S. R., 1961, *Functions of Varied Experience*, Dorsey Press, Chicago, Illinois.

Gentile, A. M., Green, S., Nieburgs, A., Schmelzer, W., and Stein, D. G., 1978, Disruption and recovery of locomotor and manipulatory behavior following cortical lesions in rats, *Behav. Biol.* **22**:417–455.

Goldberger, M. E., 1972, Restitution of function in the CNS: The pathological grasp in *Macaca mulatta, Exp. Brain Res.* **15**:79–96.

Goldberger, M. E., 1974, Recovery of movement after CNS lesions in monkeys, in: *Plasticity and Recovery of Function in the Central Nervous System* (D. G. Stein, J. J. Rosen, and N. Butters, eds.), Academic Press, New York, pp. 265–337.

Hall, T. C., Miller, A. K. H., and Corsellis, J. A. N., 1975, Variations in the human Purkinje cell population according to age and sex, *Neuropathol. Appl. Neurol.* **1**:345–367.

Honzik, C. H., 1936, The sensory basis of maze learning in rats, *Comp. Psychol. Mono.* **13**:1–113.

Jackson, J. H., 1879, On affection of speech from disease of the brain, *Brain* **2**:323–356.

Klisz, D., and Parsons, O. A., 1977, Hypotheses testing in younger and older alcoholics, *J. Stud. Alcohol.* **38**:1718–1729.

Krechevsky, I., 1932, "Hypotheses" in rats, *Psychol. Rev.* **39**:516–532.

Lashley, K. S., 1929, *Brain Mechanisms and Intelligence*, University of Chicago Press, Chicago.

Lashley, K. S., 1950, In search of the engram, in: *Symposium, Society for Experimental Biology*, Volume 4, Cambridge University Press, England, pp. 454–482.

Laurence, S., and Stein, D. G., 1978, Recovery after brain damage and the concept of localization of function, in: *Recovery from Brain Damage: Research and Theory* (S. Finger, ed.), Plenum Press, New York, pp. 369–409.

Leonard, C. M., 1969, The prefrontal cortex of the rat. I. Cortical projection of the mediodorsal nucleus. II. Efferent connections, *Brain Res.* **12**:296–320.

Levine, M., 1966, Hypothesis behavior by humans during discrimination learning, *J. Exp. Psychol.* **71**:331–338.

Meyer, D. R., Isaac, W., and Maher, B., 1958, The role of stimulation in spontaneous reorganization of visual habits, *J. Comp. Physiol. Psychol.* **51**:546–548.

Meyer, P. M., 1974, Recovery of function following lesions of the subcortex and neocortex, in: *Plasticity and Recovery of Function in the Central Nervous System* (D. G. Stein, J. J. Rosen, and N. Butters, eds.), Academic Press, New York, pp. 217–236.

Meyer, P. M., Horel, J. A., and Meyer, D. R., 1963, Effects of d-amphetamine upon placing response in neocorticate cats, *J. Comp. Physiol. Psychol.* **56**:402–404.

Mufson, E. J., and Stein, D. G., 1980, Degeneration in the spinal cord of old rats, *Exp. Neurol.* **70**:179–186.

Nonneman, A. J., and Kolb, B., 1979, Functional recovery after serial ablation of prefrontal cortex in the rat, *Physiol. Behav.* **22**:895–901.

Rosen, J. J., and Stein, D. G., 1969, Spontaneous alternation behavior in the rat, *J. Comp. Physiol. Psychol.* **68**:420–426.

Sabel, B., and Stein, D. G., 1981, Extensive loss of subcortical neurons in the aging rat brain, *Exp. Neurol.* **73**:507–516.

Scheff, S. E., Bernardo, L. S., and Cotman, C. W., 1978, Decrease in adrenergic axon sprouting in the senescent rat, *Science* **202**:775–778.

Scheibel, A. B., 1980, Aging and senescence in selected motor systems of man, in: *The Psychobiology of Aging: Problems and Perspectives* (D. G. Stein, ed.), Elsevier/North-Holland, New York, pp. 273–282.

Scheibel, M. E., and Scheibel, A. B., 1975, Structural changes in the aging brain, in: *Aging*, Volume 1 (H. Brody, D. Harman, and J. Ordy, eds.), Raven Press, New York, pp. 11–37.

Sperry, R. W., 1945, The problem of central nervous reorganization after nerve regeneration and muscle transposition, *Q. Rev. Biol.* **20:**311–369.

Sperry, R. W., 1947, Effect of crossing nerves to antagonistic limb muscles in the monkey, *Arch. Neurol. Psychiat.* **58:**452–473.

Stein, D. G., and Firl, A. C., 1976, Brain damage and reorganization of function in old age, *Exp. Neurol.* **52:**157–167.

8

Age, Brain Damage, and Behavioral Recovery

ARTHUR J. NONNEMAN, JOHN P. VICEDOMINI,
JAMES V. CORWIN, STEPHEN D. CURTIS,
and WALTER L. ISAAC

1. INTRODUCTION

Although there have been numerous anecdotal and experimental reports of diminished memory and learning capacity in aged humans and animals (cf. Birren and Schaie, 1977), the development of animal models to study cognitive changes during aging has not been easy. The detection of performance deficits associated with advanced age depends on the specific sensory, motor, or cognitive requirements of the behavioral task(s) studied (Campbell *et al.,* 1980; Doty, 1966; Goodrick, 1972; Rigter *et al.,* 1980). It also depends on the general conditions under which the subjects are housed and tested (Doty, 1968, 1972). Behavioral tasks that seem to be particularly sensitive to age-related cognitive changes in rats are tests of spatial memory (Barnes and McNaughton, 1980) and tasks that require the withholding of dominant response tendencies (Gold and McGaugh, 1975). Aged rats show a greater tendency than young adult rats to perseverate behavioral responses (Elias and Elias, 1976). This lack of "response flexibility" is particularly striking when the rats are required to avoid a location or to alter a response that previously was reinforced.

ARTHUR J. NONNEMAN, JOHN P. VICEDOMINI, JAMES V. CORWIN, STEPHEN D. CURTIS, and WALTER L. ISAAC • Department of Psychology, University of Kentucky, Lexington, Kentucky 40506. *Present address for JVC:* Department of Neurology, University of Florida Medical Center, Gainesville, Florida 32610. *Present address for SDC:* Psychology Department, Indiana University, Bloomington, Indiana 47401.

Perhaps this lack of behavioral flexibility also reduces the capacity for functional recovery after brain damage in aged rats. Several authors have reported a lack of recovery in aged rats (relative to young adult subjects) after damage to somatosensory cortex (Walbran, 1976) or frontal cortex (Mufson and Stein, 1980; Stein, 1974; Stein and Firl, 1976). Since these results are consistent with reports of behavioral and neural deterioration during senescence, it is easy to overgeneralize from them. There is a strong tendency to expect aged subjects always to show reduced recovery of function after brain damage.

The basic aim of this chapter is to show that deficient recovery is not inevitable in aged subjects suffering brain damage. However, the amount of recovery in aged subjects may be more dependent on the specific task requirements and testing conditions than is the recovery of young adults. Under some test conditions aged rats with brain damage show as much recovery as their younger counterparts. This is true even on tasks requiring spatial memory and/or response flexibility.

2. SPATIAL ALTERNATION LEARNING

2.1. *General Methodology*

One task that requires both spatial memory and response flexibility is spatial-delayed alternation in a T-maze. We have used this task extensively to study sparing and/or recovery of function after lesions of the medial prefrontal (MF) cortex in rats.

Whereas normal adult rats learn within three or four days to alternate successive responses to the right and left goal arms of the T-maze, adult rats with bilateral lesions of MF cortex (defined by Leonard, 1969) requires three or four times as long to learn this task. In contrast, rats that suffer bilateral removal of the same cortex as infants (10 days old) are not impaired. Rats sustaining lesions as juveniles (25 days old) are intermediate in performance between infant and adult operates (Nonneman and Corwin, 1981). Thus, age at surgery is a very important variable in determining the effects of medial prefrontal cortex damage on this task.

In the simplest version of this task, 0-sec delay, the hungry rat is placed in the stem of the T-maze and allowed to enter either goal arm for a food reward. The animal is then removed from the goal arm and placed into the stem of the maze for the next trial. The task can be made more difficult by placing the animal into a holding cage for an intertrial delay. In either event, the rat is required to enter the maze arm opposite its previous choice in order to receive a reward. If it enters the same arm as on the previous trial, that entry is scored as an initial error (nonperseverative). All subsequent reentries of the same arm are scored as perseverative errors. Details of this test procedure can be found in Nonneman and Kolb (1979).

2.2. Single-Stage Lesion Effects

Since the spatial alternation task requires the rat to remember which arm was chosen on the preceding trial and also requires it to enter the arm opposite the location of the previous reward, it seemed likely to us that senescent rats would have difficulty learning this task. This was also suggested by the fact that Mufson and Stein (1980) and Stein and Firl (1976) reported extensive degeneration of the frontal cortex in aged rats, and destruction of this cortex in adult rats produces serious spatial alternation deficits.

Juvenile (35 days old), adult (150 days old), and aged (570 days old) Sprague-Dawley albino male rats were purchased from Harlan Industries (Indianapolis). Half of the animals in each age group received sham lesions and half received bilateral removal of the MF cortex. (This cortex lies on the medial wall of the cerebral hemisphere dorsal to the olfactory bulb and tract and anterior and dorsal to the genu of the corpus callosum.) These subjects received 70 days of postoperative recovery before spatial alternation testing. During this time they were individually housed in standard wire mesh suspended cages.

Ten other senescent (800 days old) male Sprague-Dawley rats served as an additional group of unoperated controls. Because of the expense of these animals, their age, and the limited number available, they were not subjected to any surgical procedure. They were housed in translucent plastic box cages fitted with filter bonnets.

Figure 1 presents the results of 0-sec spatial alternation testing. Bilateral MF lesions produced equivalent deficits in all three lesion groups. Among the control groups, only the 830-day-old rats displayed a learning deficit. They required as many trials (days) to reach the learning criterion as the MF-operated rats, but they were intermediate to the MF and sham operates in perseverative and nonperseverative errors.

2.3. Serial Lesion Effect

In several previous studies, we have seen that the spatial alternation deficit produced by MF lesions in adult rats can be partially ameliorated if the lesion is produced in two stages separated by at least two weeks of interoperative recovery (e.g., Corwin et al., 1981). In an effort to explain this serial lesion effect, many authors have suggested that the successive destruction of tissue has facilitated structural and/or functional reorganization. For example, Loesche and Steward (1977) argued that the recovery of spatial alternation behavior that follows uni-

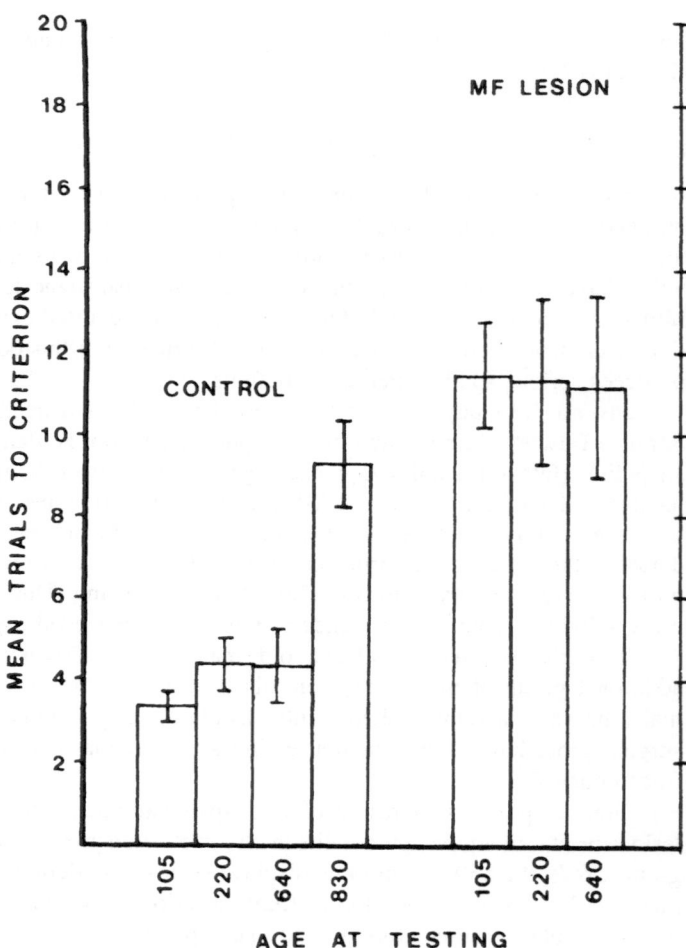

Figure 1.   Mean trials (left panel) and errors (right panel) to criterion ($\pm$ SEM) for acquisition of food-reinforced spatial alternation in a T-maze. Black portion of error bars indicates perseverative errors. Open portion of error bars indicates nonperseverative errors. MF, medial prefrontal cortex.

lateral destruction of the entorhinal cortex depends on the reinnervation of the hippocampal formation by the undamaged contralateral entorhinal cortex. In a related study, Scheff, Benardo, and Cotman (1977) reported that progressive (serial) destruction of the rat hippocampal system accelerates axonal sprouting. They suggested that this might account for the partial recovery of function allowed by serial lesion production.

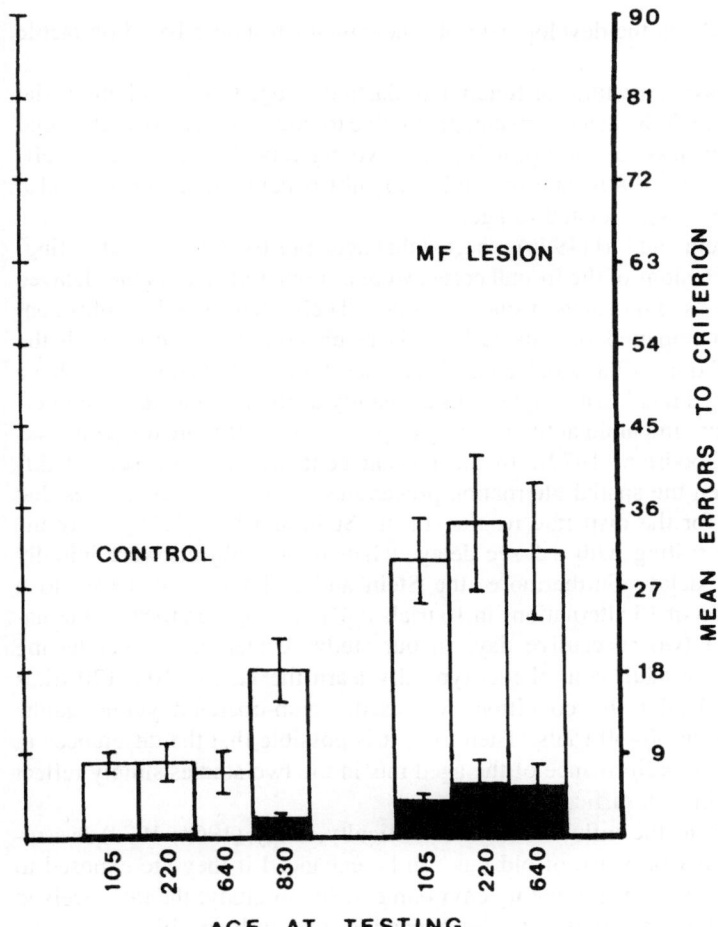

Figure 1 (continued)

Even in cases in which structural reorganization cannot be demonstrated, partial recovery might be seen if a behavior can be performed in more than one way. In this case of functional plasticity, the brain-damaged animal would accomplish a behavioral task by substituting an alternative solution. For example, Goldberger (1972) demonstrated that premotor cortex lesions in rhesus monkeys (area 6) produce a pathological grasp response. Recovery from this pathological response depends on the development of a new motor response based on tactile evasion.

In either case (structural or functional plasticity), aged rats receiving serial brain damage should be at a disadvantage relative to younger serial operates. Old rats demonstrate less axonal sprouting than young rats (Cotman and Scheff, 1979; Scheff *et al.*, 1980), and the ability to shift behavioral responses is also thought to be inversely related to age.

Indeed, Stein and Firl (1976) reported that aged rats (620 days old at testing) receiving serial lesions of the frontal cortex were as impaired on a spatial-delayed alternation task as single-stage frontal operates. Their aged (620 days old) controls were also impaired on this task. This result contrasts sharply with the performance of our 640-day-old control rats (see Fig. 1). Several studies have reported that aged rats learn simple tasks as readily as their young adult counterparts, but they are impaired at the learning or performance of more difficult tasks (Doty, 1966; Goodrich, 1972). In the present context, there are several differences between the spatial alternation procedures used in the two studies that might account for the disparate results. In the Stein and Firl (1976) study the animals started testing with a 5-sec delay, whereas our subjects were initially tested at 0-sec delay. Furthermore, the Stein and Firl rats were tested to a learning criterion of 15 alternations in 16 trials (94%) as opposed to 80% alternation on each of two successive days in our study. Under the former testing conditions, young adult control rats typically learn the task in 100–120 trials (Stein, 1974). Under the conditions we used, sham-operated young adults learned the task in 40–50 trials. Therefore, it is possible that the differences in spatial alternation performance of the aged rats in the two studies simply reflect this difference in task difficulty.

In addition to the influence of task difficulty, Doty (1968, 1972) has reported that the performance of old rats can be enhanced if they are exposed to handling or an enriched stimulating environment. In our study, the rats received extensive handling, adaptation to the apparatus, and task-specific pretraining (see Nonneman and Kolb, 1979, for a description of this procedure). It seems likely that this procedure also may have contributed to the efficient alternation learning of our aged subjects.

Because of the importance of such factors as testing conditions, task difficulty, handling, and environmental stimulation for the behavioral performance of aged control animals, it may be that similar influences are especially important for demonstrating a serial lesion recovery effect in aged rats. In other words, test conditions might be even more important for aged rats with brain damage than for aged control rats.

In an effort to test this possibility, we subjected juvenile (35 days old), adult (150 days old), and aged (570 days old) rats to serial MF cortex lesions and tested them for spatial alternation learning under conditions that were expected to enhance serial lesion recovery: long interoperative (50 days) and postoperative (21 days) recovery intervals, extensive adaptation and pretraining, task-specific interoperative experience, and intensive daily handling throughout all phases of the study. At the beginning of postoperative data collection the young, adult, and aged rats were 105, 220, and 640 days old respectively. The subjects were tested initially on a 0-sec delay task, and after they had reached the 80% alternation criterion on each of two consecutive days, they were tested with a 10-sec intertrial delay (same criterion). The initial results (Corwin et al., 1982) were quite surprising. Not only did the aged control rats perform as well as their younger counterparts, but the aged serial operates showed a serial lesion recovery effect that was indistinguishable from that of the adult serial operates.

Subsequently, we have tested additional animals under the same conditions, and the results are substantially the same. The combined results of the initial study and the subsequent replication are presented in Fig. 2. In initial testing at 0-sec delay, the aged serial operates performed significantly better than their youngest counterparts. Thus, whereas the juvenile and adult serial operates were significantly better than their age-matched single-stage operates, they were significantly worse than their age-matched sham controls on all performance measures (trials, perseverative errors, and nonperseverative errors). In contrast, the aged serial operates learned the task as readily as their age-matched sham controls. On the 10-sec delay phase, all serial operates showed equivalent serial lesion recovery effects on all measures of spatial alternation performance.

3. EMOTIONAL REACTIVITY

Even with the extensive handling, adaptation, and pretraining, our aged rats appeared to be more reactive to each change in procedure than the young adult subjects. For example, several of the aged subjects resisted handling, squealed, defecated, and urinated when they were first placed in the holding cages at the side of the T-maze during 10-sec delay testing. It should be noted that these animals had not previously been exposed to the holding cages, although they had been adapted to the T-maze itself, and they were calm during 0-sec testing. The younger subjects rarely reacted in this manner to any change.

Although all of us had the same impression that the old rats were hyperreactive, it was not possible to quantify this impression during alternation testing. Therefore, an additional group of 540-day-old and 180-day-old unoperated Harlan Sprague-Dawley rats were rated for emotional reactivity using a modified version (Nonneman et al., 1974) of the emotionality rating scale described by King (1958). The rats were rated on a five-point scale for reactiveness to a pencil positioned 1 cm in front of the animal's snout, reactions to a light tap on the back with a pencil tip, resistance to capture by a gloved hand, resistance to handling,

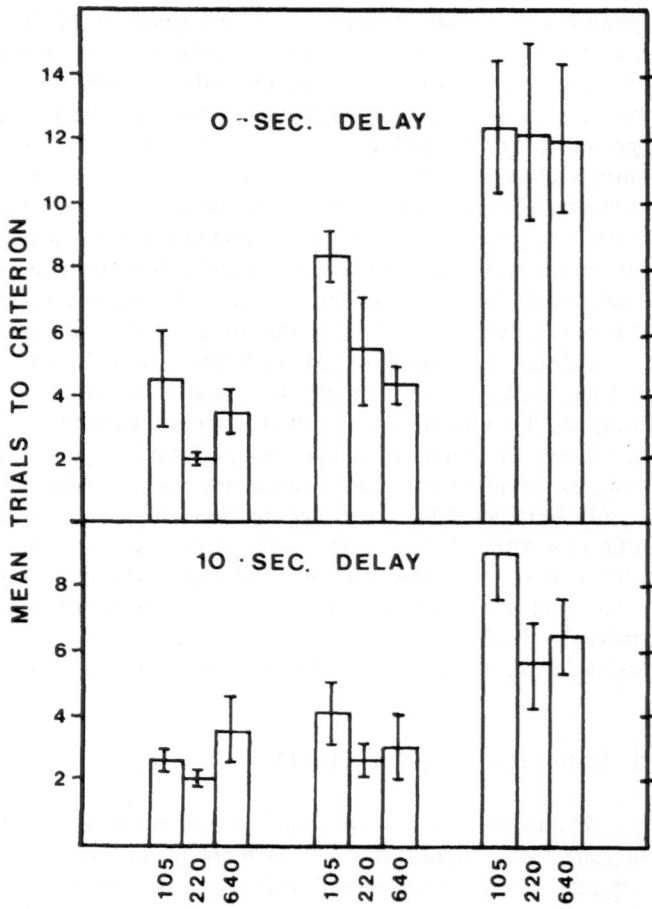

Figure 2.   Mean trials (left panels) and errors (right panels) to criterion (± SEM) for acquisition of spatial-delayed alternation at 0-sec delay (upper panels) and 10-sec delay (lower panels) intertrial intervals. Within each panel left bars represent sham-operated controls, center bars represent serial

and vocalization during testing. The results of this experiment are presented in Fig. 3.

On days, 1, 3, and 4, the aged rats were significantly more reactive than the young adults. However, both age groups habituated to the test stimuli at an equal rate. In other words, there was no interaction between age and test sessions. This confirms our subjective impression that aged rats are more reactive emotionally than young rats; but it is also consistent with our impression that the hyperreactivity of the aged rat can be reduced with sufficient handling and adaptation to the

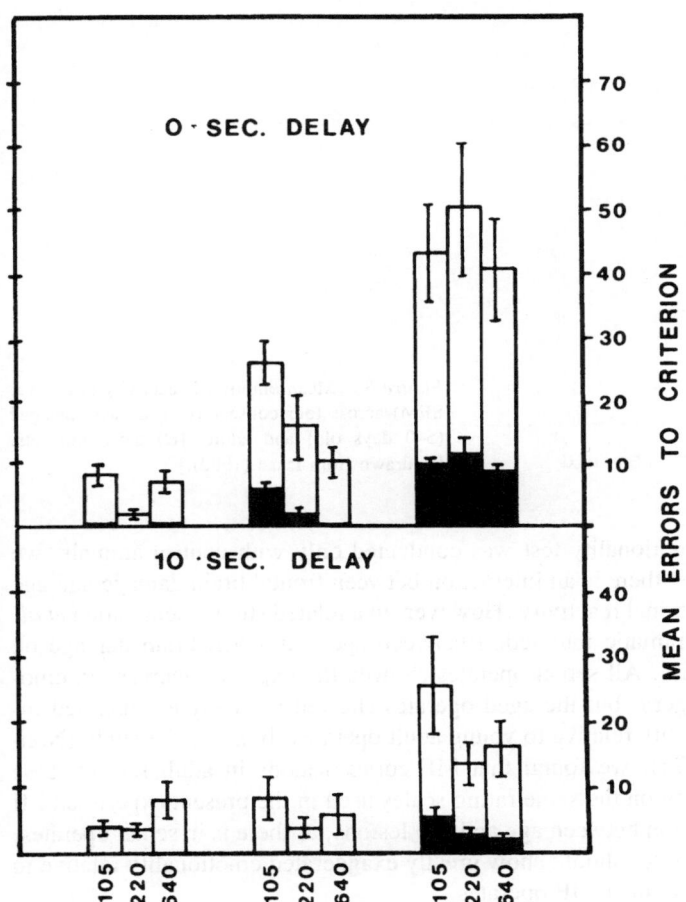

MF operates, and right bars represent single-stage MF operates. Black portion of error bars indicates perseverative errors, and open portion of error bars indicates nonperseverative errors. MF, medial prefrontal cortex.

testing situation. It is worth noting that the two age groups differed primarily in resistance to capture and handling. It should also be noted that all of our subjects in the two studies mentioned above received at least five days of handling and adaptation to the test situation before data collection. It seems likely that emotional reactivity would interfere with initial learning in most appetitively reinforced learning tasks. If we had tested our animals without adaptation or pretraining, our aged rats might have required more trials (and made more errors) to reach the learning criterion than their younger counterparts.

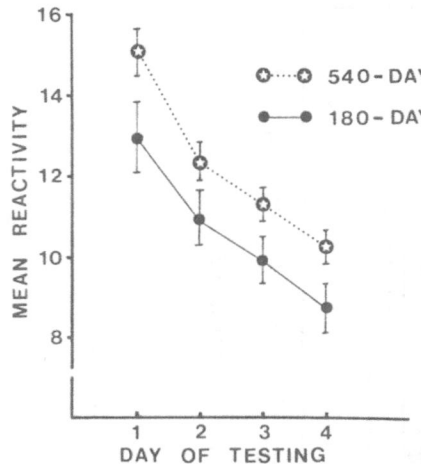

Figure 3. Mean emotional reactivity scores (± SEM) across four consecutive test days for aged (540 days old) and adult (180 days old) rats. [Redrawn from Isaac (1982).]

Since this emotionality test was conducted only with control animals, we cannot say whether there is an interaction between frontal brain damage and age of subject on emotional reactivity. However, in a related study, Bengelloun *et al.* (1980) found a dramatic interaction between age and septal brain damage on emotional reactivity. All septal operates showed the expected increase in emotionality after surgery, but the aged operates showed a greatly exaggerated increase in emotionality relative to young adult operates. In an earlier study (Nonneman *et al.,* 1974), we found that MF cortex lesions in adult rats increase emotional reactivity on the same rating scales used in the present experiment. If there is an interaction between age and MF lesions, as there is in septal operates, then aged MF operates should show greatly exaggerated emotionality relative to aged controls or younger MF operates.

Bengelloun *et al.* (1980) also found an interaction between age and septal damage on activity in a novel open field but not in the familiar home cage. The aged septal rats froze more and stood up (reared) less than any other group in the open field. Several other authors, including ourselves, have noted that aged subjects are more immobile in maze-learning situations than young subjects, and this immobility can be a serious confound in the assessment of the cognitive ability of aged subjects with or without brain damage.

4. DRL-20 PERFORMANCE

If we are correct that the appearance of a serial lesion recovery effect in aged animals depends on test conditions that minimize emotional reactivity and locomotor "freezing," then operant-testing techniques that eliminate the need for intertrial handling, minimize external distractions, and do not require a loco-motor response should allow aged brain-damaged rats to demonstrate their true cognitive abilities. In a previous study, Curtis and Nonneman (1977) examined the effects of single-stage and serial hippocampal destruction on differential reinforcement of low response rate (DRL-20) performance in adult rats. Food-deprived rats were first shaped to press a bar on a continuous reinforcement (CRF) schedule for a food reward. After ten days of CRF training the rats were shifted to an unsignaled DRL-20 schedule that required them to wait at least 20 sec between responses in order to receive food reinforcement. This was initially difficult even for unoperated rats, but they became quite proficient after a month of daily testing. In contrast, rats with single-stage hippocampal lesions seemed to be incapable of withholding responses. They perseverated the high-response rate characteristic of their earlier CRF training and showed no improvement in re-sponse efficiency (total reinforcements/total responses) even after two months of daily testing. The rats receiving two-stage hippocampal lesions were intermedi-ate in performance. They showed a significant but incomplete serial lesion recov-ery effect.

In order to determine whether this serial lesion recovery might be age-dependent, we decided to replicate this study with three age groups: juvenile (35 days old), adult (160 days old), and aged (570 days old) at the time of first surgery. It seemed quite possible that the aged controls would have difficulty with this task and in addition that the aged serial operates might be as impaired as single-stage operates. A number of papers have reported degenerative changes in the hippocampus of aged rats (Bondareff, 1981; Bondareff and Geinisman, 1976; Lamar et al., 1976). On the other hand, a serial lesion effect might be evident even in the senescent rats if the elimination of emotional hyperreactivity and locomotor inactivity were critical.

All rats received two operations separated by 50 days. Control rats received sham lesions at stage one, and rats with single-stage hippocampal destruction (group H) received electrolytic lesions of the entire hippocampus at first surgery. There were two serial lesion groups: group D-V received a lesion of the dorsal component of the hippocampus at first surgery and a lesion of the ventral compo-nent at second surgery, whereas group V-D received ventral and dorsal lesions at first and second surgery, respectively. During the 50-day interoperative interval, all subjects received five days of CRF testing followed by 40 days of DRL-20 testing. Following three days of postoperative recovery after the second stage of surgery, they then received an additional 30 days of DRL-20 testing.

The results are shown in Fig. 4. During the 40 days of interoperative training, the rats with only the ventral component of the hippocampus destroyed were intermediate in performance to the controls and the rats with the entire hippocampus destroyed. Rats with only the dorsal component destroyed were equivalent to controls, and rats with total hippocampus destruction differed from all other groups. The same pattern was observed in all age groups.

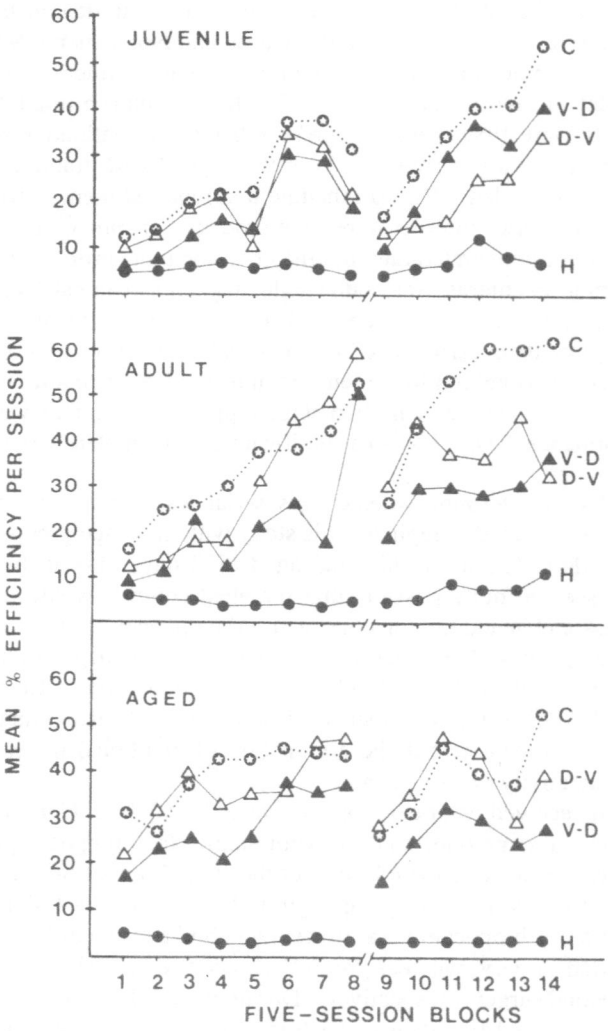

Figure 4.   Mean DRL-20 efficiency for juvenile, adult, and aged rats. At the time of first surgery, the juvenile rats were 35 days old, the adult rats were 160 days old, and the aged rats were 570 days old. Group C (control) rats received sham lesions at first surgery. Group D-V rats received lesions of the dorsal and ventral components of the hippocampus at first and second surgery, respectively. Group V-D rats received lesions of the ventral and dorsal components of the hippocampus at first and second surgery, respectively. Group H rats received a lesion of the entire hippocampus at first surgery. The break in the abscissa indicates the time of the second surgery. [Redrawn from Curtis (1979).]

During the 30 days of postoperative DRL testing, a similar pattern was seen. There were no significant age effects or interactions. However, within each age the serial operates were intermediate in performance to the control and single-stage operates, and groups C and H differed from all other groups. Thus, there was a significant and equivalent serial lesion recovery effect at all ages. This serial lesion effect cannot be attributed to differences in lesion size between single-stage and serial operates, since lesion size was equivalent (almost total hippocampal destruction) for all lesion groups at all ages.

5. DISCUSSION

In previous research with aged rats, a number of authors have noted that the performance of aged subjects is very much dependent upon the specific behavioral requirements of the task(s) under study. In many cases, the performance of aged but otherwise normal rats seems strikingly normal. Under other conditions aged animals may show clear-cut disabilities (e.g., Campbell *et al.*, 1980; Rigter *et al.*, 1980). Based on the results of these experiments, the same seems to be true of recovery from brain damage in aged rats. Despite reports of declining neural and behavioral plasticity in aged rats (cited earlier), we did not see any decline in the magnitude of the serial lesion recovery effect after lesions of medial prefrontal cortex or hippocampus. Thus, it appears that a loss of recovery potential with advancing age may not be inevitable. However, it is important to note that these results were obtained under conditions that were carefully chosen to enhance the strength of a serial lesion effect. If we are correct that aged subjects are more dependent than young adults on favorable test conditions, then one should expect different results under less optimal circumstances. Specifically, the results of these studies suggest that it is important to test the aged subjects under conditions that minimize emotional reactivity if one hopes to obtain an accurate index of their learning capacity and/or recovery potential.

We should also note that our use of the word *recovery* is not intended to imply that our serial lesion subjects at any age are solving the task in exactly the same way that a normal animal would. We cannot make such a determination with these data. We can only determine whether the subjects reached the same performance level or learned at the same rate. Determining exactly how the animals have performed the task is much more difficult; but ultimately it is extremely important to determine whether aged subjects with or without brain damage have performed in the same manner as younger subjects in the same situation. Therefore, it is encouraging to see that a start has been made in this direction (Valentino and Stein, Chapter 7).

ACKNOWLEDGMENTS. Portions of the data reported here are contained in a master's thesis by W. L. Isaac and in doctoral dissertations by J. V. Corwin and

S. D. Curtis submitted to the Graduate School of the University of Kentucky. This research was supported by grants from the University of Kentucky Research Foundation, the Sanders-Brown Kentucky Research Center on Aging, the University of Kentucky Graduate School, Sigma Xi, and Biomedical Science Research Grant 2-S07-RR07114. The authors would like to thank Linda Ristine, Lisa Tolnitch, and Sarah Cox for assistance with behavioral testing.

REFERENCES

Barnes, C. A., and McNaughton, B. L., 1980, Spatial memory and hippocampal synaptic plasticity in senescent and middle-aged rats, in: *The Psychobiology of Aging: Problems and Perspectives* (D. G. Stein, ed.), Elsevier/North-Holland, New York, pp. 253–272.

Bengelloun, W. A., El Hilali, M., Bouizzar, Z., and Viellat, J. P., 1980, Brain damage in mature and aged rats: Behavioral effects of septal lesions, in: *The Psychobiology of Aging: Problems and Perspectives* (D. G. Stein, ed.), Elsevier/North-Holland, New York, pp. 253–272.

Birren, J. W., and Schaie, K. W., 1977, *Handbook of the Psychology of Aging*, Van Nostrand, New York.

Bondareff, W., 1981, The neurobiological basis of age-related changes in neuronal connectivity, in: *Aging: Biology and Behavior* (J. L. McGaugh and S. B. Kiesler, eds.), Academic Press, New York, pp. 141–158.

Bondareff, W., and Geinisman, Y., 1976, Loss of synapses in the dentate gyrus of the senescent rat, *Am. J. Anat.* **145:**129–136.

Campbell, B. A., Krauter, E. E., and Wallace, J. E., 1980, Animal models of aging: Sensory-motor and cognitive function in the aged rat, in: *The Psychobiology of Aging: Problems and Perspectives* (D. G. Stein, ed.), Elsevier/North-Holland, New York, pp. 201–226.

Corwin, J. V., Nonneman, A. J., and Goodlett, C., 1981, Limited sparing of function on spatial delayed alternation after two-stage lesions of prefrontal cortex in the rat, *Physiol. Behav.* **26:**763–771.

Corwin, J. V., Vicedomini, J. P., Nonneman, A. J., and Valentino, L., 1982, Serial lesion effect in rat medial frontal cortex as a function of age, *Neurobiol. Aging* **3:**69–76.

Cotman, C. W., and Scheff, S., 1979, Synaptic growth in aged animals, in: *Physiology and Cell Biology of Aging*, Volume 8 (A. Cherkin, C. Finch, T. Makinodan, F. Scott, B. Strehler, and N. Kharasch, eds.), Raven Press, New York, pp. 109–120.

Curtis, S. D., 1979, The effects of bilateral seriatim versus single-stage hippocampectomy upon the acquisition of a DRL20 operant schedule in juvenile, adult and aged rats, doctoral dissertation, University of Kentucky.

Curtis, S. D., and Nonneman, A. J., 1977, The effects of successive bilateral hippocampectomy on DRL20 performance in the rat, *Physiol. Behav.* **19:**707–712.

Doty, B. A., 1966, Age differences in avoidance conditioning as a function of distribution of trials and task difficulty, *Genet. Psychol.* **109:**249–254.

Doty, B. A., 1968, Effects of handling on learning of young and aged rats, *J. Gerontol.* **23:**142–144.

Doty, B. A., 1972, The effects of cage environment upon avoidance responding of aged rats, *J. Gerontol.* **27:**358–360.

Elias, P. K., and Elias, M. F., 1976, Effects of aging on learning ability: Contributions from the animal literature, *Exp. Aging Res.* **2:**165–186.

Gold, P. E., and McGaugh, J. L., 1975, Changes in learning and memory during aging, in: *Neurology of Aging* (J. M. Ordy and K. R. Brizzee, eds.), Plenum Press, New York, pp. 145–158.

Goldberger, 1972, Restitution of function in the CNS: The pathologic grasp in macaca mulatta, *Exp. Brain Res.* **15:**79–96.

Goodrick, C. L., 1972, Learning by mature young and aged Wistar albino rats as a function of test complexity, *J. Gerontol.* **27:**353–357.

Isaac, W. L., 1982, Serial lesion effect in medial frontal cortex in adult and senescent rats, master's thesis, University of Kentucky.

King, F. A., 1958, Effects of septal and amygdaloid lesions on emotional behavior and conditioned avoidance responses in the rat, *J. Nerv. Ment. Dis.* **126:**57–63.

Lamar, C. H., Hinsman, E. J., and Henrikson, C. K., 1976, Alterations in the hippocampus of aged mice, *Acta Neuropathol.* **36:**387–391.

Leonard, C. M., 1969, The prefrontal cortex of the rat. I. Cortical projections of the mediodorsal nucleus. II. Efferent connections, *Brain Res.* **12:**296–320.

Loesche, J., and Steward, O., 1977, Behavioral correlates of denervation and reinnervation of the hippocampal formation of the rat: Recovery of alternation performance following unilateral entorhinal cortex lesions, *Brain Res. Bull.* **2:**31–39.

Mufson, E. J., and Stein, D. G., 1980, Behavioral and morphological aspects of aging: An analysis of rat frontal cortex, in: *The Psychobiology of Aging: Problems and Perspectives* (D. G. Stein, ed.), Elsevier/North-Holland, New York, pp. 99–125.

Nonneman, A. J., and Corwin, J. V., 1981, Differential effects of prefrontal cortex ablation in neonatal, juvenile and young adult rats, *J. Comp. Physiol. Psychol.* **95:**588–602.

Nonneman, A. J., and Kolb, B., 1979, Functional recovery after serial ablation of prefrontal cortex in the rat, *Physiol. Behav.* **22:**895–901.

Nonneman, A. J., Voigt, J., and Kolb, B., 1974, Comparisons of the behavioral effects of hippocampal and prefrontal cortex lesions in the rat, *J. Comp. Physiol. Psychol.* **87:**249–260.

Rigter, H., Martinex, J. L., Jr., and Crabbe, J. C., Jr., 1980, Forgetting and other behavioral manifestations of aging, in: *The Psychobiology of Aging: Problems and Perspectives* (D. G. Stein, ed.), Elsevier/North-Holland, New York, pp. 161–175.

Scheff, S., Benardo, L., and Cotman, C., 1977, Progressive brain damage accelerates axon sprouting in the rat, *Science* **197:**795–797.

Scheff, S. W., Benardo, L. S., and Cotman, C. W., 1980, Decline in reactive fiber growth in the dentate gyrus of aged rats compared to young adult rats following entorhinal cortex removal, *Brain Res.* **199:**21–38.

Stein, D. G., 1974, Some variables influencing recovery of function after central nervous system lesions in the rat, in: *Plasticity and Recovery of Function in the Central Nervous System* (D. G. Stein, J. J. Rosen, and N. Butters, eds.), Academic Press, New York, pp. 373–427.

Stein, D. G., and Firl, A. C., 1976, Brain damage and reorganization of function in old age, *Exp. Neurol.* **52:**157–167.

Walbran, B., 1976, Age and serial ablations of somotosensory cortex in the rat, *Physiol. Behav.* **17:**13–17.

9

Age and Recovery from Brain Damage
A Review of Clinical Studies

HARVEY S. LEVIN, LINDA EWING-COBBS,
and ARTHUR L. BENTON

1. FUNCTIONAL PLASTICITY AFTER EARLY BRAIN INJURY

1.1. Implications of Ablation Experiments for Recovery in Children: Methodological Issues in Recovery from Brain Injury at Different Ages

More than a century has passed since Cotard (1868) reported that early or congenital left-hemisphere damage does not lead to aphasia. Confirmatory observations showing that aphasia acquired after a child has learned speech is less severe and less permanent than in adults contributed to the widely held view that the young brain is more resilient to insults than the mature brain (cf. Guttman, 1942). However, this view has recently come under criticism for being accepted as "unquestioned dogma" (St. James-Roberts, 1979). Focusing on age at injury as an explanatory variable de-emphasizes a number of equally important interactive variables such as the functional maturation of the involved tissue, age at assessment, testing procedures, and experiential factors related to both the preinsult and recovery periods (St. James-Roberts, 1979).

Comparison of recovery from brain injury at different ages is complicated by several factors. Differences in the pathophysiology of brain injury (e.g., frequent cerebral swelling and infrequent intracranial hematoma after head injury

HARVEY S. LEVIN • Division of Neurosurgery, The University of Texas Medical Branch, Galveston, Texas 77550.   LINDA EWING-COBBS • Department of Psychology, University of Houston, Houston, Texas 77004.   ARTHUR L. BENTON • Departments of Neurology and Psychology, University of Iowa, Iowa City, Iowa 52240.

in children) complicate efforts to equate these injuries using a scale of severity. When investigators consider the long-term effects of injury sustained in infancy, the chronicity of brain damage is typically greater than when brain injury occurs at a later age. The variety of etiologies producing focal and diffuse brain injury also complicates analysis of the effect of age on recovery. Moreover, these etiologies are not uniformly distributed across the lifespan (e.g., focal disruption of the developing brain by cerebrovascular disease and cortical tumors is uncommon while meningitis is a relatively frequent cause of diffuse insult to the young brain). In contrast to animal experiments which may control for the locus and extent of lesion, prelesion training, and the postlesion interval before retesting, clinical investigators can rarely control variables other than the injury–test interval and they can only approximately match groups with respect to the locus of lesion. As will be seen, however, the results of experimental studies and the clinical findings converge in disputing a simplistic view of cerebral plasticity.

Early experiments carried out by Kennard (1938) supported the concept of cerebral plasticity. She reported sparing of severe motor deficit immediately after ablation of the motor cortex in infant monkeys, while corresponding lesions in mature animals produced severe impairment. Kennard observed that as the operated infants matured, however, they exhibited motor incoordination and spasticity. Sparing of function after brain ablation, which has also been demonstrated in the infant visual and somatosensory systems (Eidelberg and Stein, 1974; Finger, 1978), is reviewed in other chapters of this book. Nevertheless, limitations in plasticity of function have become increasingly evident.

Goldman (1974) demonstrated that orbitofrontal lesions produce comparable deficits in both young and mature animals on a delayed alternation task, whereas there was apparent sparing after dorsolateral frontal ablation during infancy. She observed a late deficit in animals that had been subjected to dorsolateral lesions as infants, suggesting a variable rate of development among different regions of frontal cortex. Goldman and Alexander (1977) have observed similar performance deficits by using reversible cryogenic depression to temporarily inactivate dorsolateral tissue in normally developing animals. It may be inferred from Goldman's work that the encephalization of neural structures must be considered when interpreting lesion studies since the relative maturity of a given substrate influences the amount of recovery of sparing of function observed after brain injury. Furthermore, the developmental course of recovery can be examined only within a longitudinal paradigm.

The results of animal experiments have been informative but even these studies are frequently compromised by methodological problems. The duration of the postlesion period has not been controlled in some studies, thereby giving the adults less time to recover. Even when duration of recovery has been held constant, the experiential component between groups differs significantly. Other studies have not included mock operations to expose controls to the trauma of craniotomy and possible postoperative infection. Assessment of behavioral outcome is typically restricted to one or two tasks. Data on the modification of performance on these tasks over time is lacking as is knowledge about encephalization of specific brain regions believed to mediate performance at different developmental stages. Carefully controlled longitudinal studies that examine a variety of behaviors over a significant portion of the lifespan are required are for a complete assessment of delayed deficits.

1.2. Perinatal and Neonatal Injury

1.2.1. Pathophysiology

Knobloch and Pasamanick (1959) postulated a "continuum of reproductive casualty," encompassing a wide range of outcomes varying from behavioral disturbance to mental retardation and depending on the extent and site of perinatal or neonatal brain damage. The two major etiologies of early brain damage are anoxia (deprivation of oxygen supply) and hypoxia (reduction in oxygen supply). Although neuropathologic studies suggest that the fetal brain is damaged by loss of oxygen, animal experiments suggest that neonates are more resistant to hypoxia−ischemia than adults (Duffy et al., 1982).

In comparison with adults sustaining injury of a similar etiology, the locus and type of brain injury vary in the fetus (Norman, 1978; Towbin, 1969). Hypoxia−ischemia in the perinatal period leads to cell loss in the cerebral cortex, thalamus, and basal ganglia. There is also focal damage to brain stem nuclei and involvement of hemispheric white matter. In contrast, similar insult in adults is mainly confined to gray matter structures, with the hippocampus, cerebellar cortex, and arterial boundary zones of the cerebral cortex being the most vulnerable regions.

Brain injury detected during the prenatal period also may originate in an intrauterine pathologic process such as infection of the placenta. Hypoxia is the most common type of brain damage in the premature newborn with a predilection for the germinal matrix deep in the cerebrum and the periventricular white matter. The watershed zone between the anterior and middle cerebral arteries is also vulnerable to hypoxic injury. Cerebral palsy is a frequent sequel to hypoxic injury in the premature fetus, whereas hypoxic injury in the term newborn more frequently produces cortical injury manifested acutely by seizures and later by mental retardation. Mechanical injury to the brain (e.g., subdural hematoma) occurs primarily when the fetus is expressed through the birth canal. The brain stem is particularly vulnerable in these cases.

In comparing the prognosis for recovery from hypoxic−ischemic injury in infants to that of adults following stroke, Towbin (1969) believes that neonates exhibit more impressive recovery during the early postinjury period. However, he cites cystic scarring and a loss of "building material" of the immature brain as factors that contribute to delayed sequelae of neonatal injury.

1.2.2. Residual Neuropsychological Functioning after Perinatal/Prenatal Injury

Graham and her colleagues (1962) found that three-year-olds who had sustained perinatal anoxia were cognitively impaired in comparison with a control group. The results of this prospective study were also compatible with the view of anoxia as a continuum rather than an "all-or-none" event.

Rudel, Teuber, and Twitchell (1974) studied 63 children (age range, 7–18 years) of whom 59 had perinatal or prenatal brain damage. Although two thirds of the series had IQs below 80, there was marked variability. Complex motor deficit (e.g., decreased alternating movements, synkinesia) was frequently present irrespective of the child's IQ. Sensory defects were rare and confined to diminished response to double-simultaneous stimulation. Oculomotor disturbances (e.g., impaired tracking, eye–head incoordination) were common as were disturbances of the body schema and impairment in spatial thinking. The adult pattern of lateralized neuropsychological impairment was present in children with predominantly right-hemisphere injury, whereas children who sustained left-hemisphere damage escaped from unequivocal aphasia. In an earlier study Rudel and Teuber (1971) found that defective route finding by reference to a visual map was associated with a low performance IQ score whether or not language scores were relatively high. The impairment of the brain-damaged children cannot be attributed entirely to global mental deficit because the authors matched these patients to control subjects on the basis of mental age.

1.3. Early Lateralized Brain Injury

1.3.1. Left Hemisphere Injury and Acquisition of Speech

Much of the evidence for the invulnerability of the immature brain to unilateral insult has been derived from studies of acquired aphasia. Support for this position was bolstered by Basser's (1962) finding of no significant differences in either verbal IQ or the onset of language following infantile hemiplegia affecting the right versus the left side. In comparison with the typical pattern found in adults with left-hemisphere injury, Basser found a transient language deficit ensued in a high proportion of left hemiplegics and that persistent language dysfunction was rare in right hemiplegics. Basser's study has been criticized by Satz and Bullard-Bates (1981) on the ground that half of the cases were either grossly mentally defective or untestable. Additionally, Basser (1962) used only crude measures of linguistic capability (e.g., age of speech onset) and failed to examine developmental aspects of language acquisition such as word combinations and syntax (Dennis and Whitaker, 1976).

Lenneberg (1967) incorporated Basser's findings into his theory of hemispheric equipotential1ty. He postulated that equipotentiality exists until the onset of speech when lateralization of function progresses, i.e., one hemisphere subsumes language functions. While injury during infancy does not impair language development since the left and right hemispheres provide equally good substrates for language development, dominant hemispheric insult at a successively later age after the onset of speech progressively results in symptoms which resemble adult aphasia. According to Lenneberg, by midadolescence the adult pattern of symptoms and recovery characteristics is established.

Neither the equipotentiality nor the progressive lateralization hypotheses have received consistent support at the anatomical or behavioral levels. Anatomical differences exist between the right and left hemispheres at birth; areas subserving language functions in the left hemisphere are asymmetric when compared to the homologous right hemisphere regions (Wada, 1974; Witelson and Pallie, 1973). Furthermore, more recent studies of infantile hemiplegics affected before the onset of speech suggest that left-hemisphere insult is associated with more linguistic deficits than comparable right-hemisphere injury (Annett, 1973; Bishop, 1981).

1.3.2. Acquired Aphasia in Children

Acquired aphasia refers to language impairment which occurs after language has developed normally. Historically, acquired aphasia in children following dominant hemisphere insult has been reported to be rare when compared with adults. However, in a critical review of this topic, Satz and Bullard-Bates (1981) concluded that acquired aphasia is not rare in children if the lesion is unilateral and involves the language areas. After infancy, the risk of aphasia following left-hemisphere injury is greater than the risk following right-hemisphere injury. Furthermore, the risk of aphasia is comparable in right-handed children and adults following left-hemisphere insult. These findings yield little support for the equipotentiality hypothesis.

Acquired aphasia in children, however, is associated with different symptoms and recovery characteristics than seen in adult aphasias. Aphasic children tend to be nonfluent and are often mute during the acute recovery period. Paraphasia and logorrhea are rare. As recovery progresses, dysarthria, hesitant speech, and a decreased lexical stock are evident following both frontal and parietotemporal lesions. Temporal lesions can result in receptive deficit in a subgroup of the cases (Hécaen, 1976; Alajouanine and Lhermitte, 1965; Guttman, 1942). The pattern of deficits appears to vary somewhat with age at injury.

Injuries sustained in older children and adolescents produce a symptom picture
that is more characteristic of adult aphasias.

Recovery in children has long been noted as being quite rapid. However,
although spontaneous recovery is marked in children, disorders of writing (Hécaen, 1976) and impaired comprehension of written material (Alajouanine and
Lhermitte, 1965) tend to persist. Estimates of recovery range from 50% (van
Dongen and Loonen, 1977) to 75% (Alajouanine and Lhermitte, 1965) to nearly
100% (Basser, 1962). Although most studies do not report recovery figures as a
function of age, there does not appear to be a linear relationship between age at
injury and subsequent language status (van Dongen and Loonen, 1977; Woods
and Teuber, 1978). Conversely, Woods and Carey (1979) observed that recovery
was best following left-hemisphere lesions acquired during infancy, intermediate
for children receiving lesions before eight years and less complete following
insult in older children. Although left-hemisphere lesions sustained before one
year of age did not yield significant linguistic dysfunction upon examination
during adolescence, impairments in higher cognitive functions were noted in
both early and late lesion groups (Woods and Carey, 1979). Even if specific
aphasic symptoms resolve following dominant hemisphere injury in children,
cognitive deficits persist regardless of age at injury (Alajouanine and Lhermitte,
1965; Hécaen, 1976; Woods and Teuber, 1978).

1.3.3. Intellectual Ability after Early Lateralized Injury

The effects of early lateralized brain injury on cognitive functioning depend
upon the age at injury. In a long-term follow-up study of patients who had
sustained prenatal or perinatal brain injury as compared with postnatal lateralized
injury (e.g., thrombosis), Woods (1980) defined an early lesion based on the
appearance of a hemiparesis before the age of one year. As depicted in Fig. 1, he
found that children with a lesion of either hemisphere before their first birthday
had verbal and performance scores below the mean expected for the normal
population. When a left-hemisphere injury occurred after age one, both verbal
and performance IQ scores were lower. Late-occurring right-hemisphere lesions
had adverse effects confined to the performance IQ.

1.3.4. Effects of Hemispherectomy on Language and Cognition in Children

There is a prevailing view that recovery after hemispherectomy is inversely
related to the age at onset of brain damage. As will be seen, this generalization
applies primarily to linguistic competence after hemispherectomy. Since post-
surgical survival is longer in infantile and juvenile cases, there is a greater
opportunity to assess cognitive recovery as compared with the relatively few

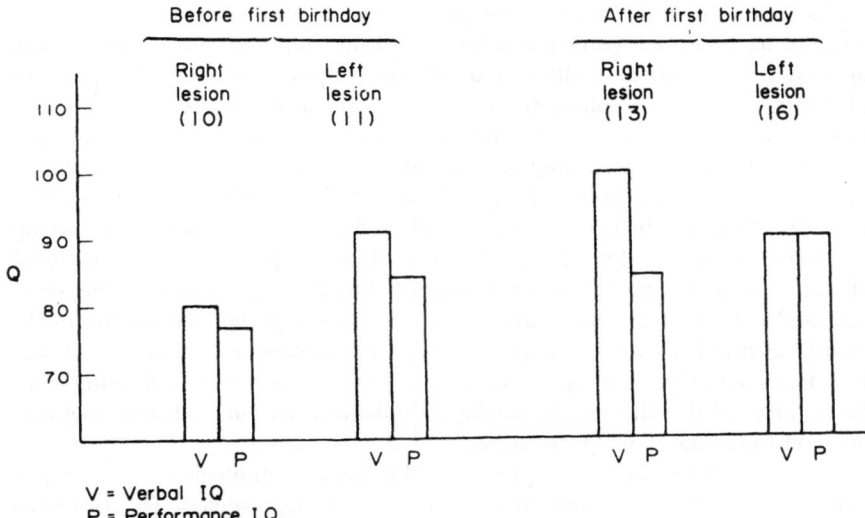

V = Verbal IQ
P = Performance IQ

Figure 1. Comparison of Wechsler Verbal and Performance IQs by side of lesion and age of occurrence for all hemiparetic patients. [From Woods (1980). Reproduced with permission of the author and publisher.]

adults who undergo hemispherectomy as a last resort because of a malignant brain tumor.

Recent reviewers (cf. St. James-Roberts, 1979) of the literature on hemispherectomy have pointed out the methodological deficiencies that detract from most of the early outcome studies. These problems include the frequent finding of severe mental deficiency which may hinder speech development, a lack of quantitative assessment of language, and inadequate statistical analysis. Notwithstanding these constraints, there is agreement that infantile left hemispherectomy is compatible with normal language development. Although there are less abundant outcome data for children who undergo left hemispherectomy after acquiring speech, it is clear that at least partial recovery of language is possible in these cases. It may be inferred that the intact right hemisphere is capable of supporting language dev.elopment during infancy and early childhood and may even assume this role more adequately after removal of an epileptogenic left hemisphere. Apart from these criticisms, the generality of the findings to hemisperic specialization of the normal brain is also open to question.

1.3.4a. Language. In Basser's (1962) series of hemispherectomized infants and children, 25 were brain damaged (12 left, 13 right) before the onset of speech. The remaining ten cases became hemiplegic after language acquisition. Postoperatively, verbal IQ and speech functions were generally unchanged and similar regardless of the side of hemispherectomy. No evidence of dysphasia was found in the early left-hemisphere lesions group (improvement in speech was reported in some patients), although dysphasic symptoms were noted in one case following late left hemispherectomy. By comparison, Basser described 20 children (13 left-brain damaged, 7 right-brain damaged) who developed hemiplegia or underwent hemispherectomy after the onset of speech. The group included three cases of left hemispherectomy and three patients with right hemispherectomy. Basser found that speech was greatly reduced or abolished for a varying duration regardless of the side of hemiplegia. Loss of speech was more frequent when the unilateral brain damage occurred before three years of age. The presence and degree of receptive impairment was not reported. Given the methodological problems discussed previously, no firm conclusions may be drawn from these data regarding age at insult and subsequent cognitive development. Basser inferred that the side of hemispheric damage was unrelated to linguistic outcome following hemispherectomy.

McFie (1961) reported dysphasia in 14 infantile hemiplegics, with an equal number of left and right hemisphere-damaged cases represented. The dysphasia was characterized primarily by errors in naming but occasionally included receptive defects. The severity of the dysphasia was not assessed, and it is unclear at what time postoperatively language functions were evaluated. Dysphasia following left-hemisphere injury was equally common in infantile and child cases, while dysphasia after right-hemisphere injury occurred almost exclusively after infantile injuries. Wilson (1970) described outcome measures on McFie's cases as well as on 16 additional cases. Speech functions were unimpaired in 84% irrespective of side of hemispherectomy. Dysphasia was noted in six cases, five of whom received left hemispherectomies. In half of these cases the loss of speech was permanent and was associated with gross mental defect.

Language dysfunction following left hemispherectomy for a tumor in a 10-year-old was described by Gott (1973a). Severe expressive and moderate receptive aphasia were observed immediately after surgery. Two years postoperatively, auditory comprehension was still relatively spared and she could give the appropriate "yes" or "no" answer to propositional statements (e.g., "elephants can read"). Spontaneous speech was, however, limited to single words or short phrases; vocabulary was diminished and dysnomia was present. Proficiency in spelling, reading, and writing was below preoperative levels. Written language was particularly distorted.

Day and Ulatowska (1979) examined an infantile hemiplegic hemispherectomized at four years of age. Aphasic symptoms were not present after a recovery period of eight years. Expressive language was seemingly normal although its content was impoverished due to a decrease in propositional speech. Deficits were apparent in grammar and in the comprehension of verb tenses, comparatives, and irregular plurals.

In contrast to the deficits mentioned above, Smith and Sugar (1975) reported a case who developed a right hemiparesis at five months of age and underwent left hemispherectomy five years later, who was evaluated when he was 26 years old. Speech, comprehension, reading, and writing were within normal limits and the verbal IQ was 126.

1.3.4b. Visuospatial. Kohn and Dennis (1974) reported that infantile left hemiplegics with seizures who underwent right hemispherectomy (from 1 to 17 years) performed adequately on simple route-finding and directional tasks. However, performance declined as task complexity increased with evident difficulties in visual perception and visual–motor integration following right hemidecortication. Impairments in visual–motor capability and constructional praxis have also been found after left hemidecortication (Day and Ulatowska, 1979). Gott (1973b) reported that both right and left hemispherectomy cases had difficulty with visual–motor integration. The left case was additionally deficient on visual organization and mazes which reflects a lack of analytic capability. Damasio, Lima, and Damasio (1975) reported limited deficits on nonverbal tasks after right hemispherectomy at five years of age. While perception of tactile direction and production of melodies was impaired, performance on mazes, route-finding tasks, facial recognition, and constructional tasks was within normal limits.

1.3.4c. Memory. Gott (1973b) suggested that memory deficits predominate after removal of either hemisphere in children. On the Wechsler Memory Scale (WMS) a left hemidecorticate exhibited several verbal memory deficits superimposed on generally depressed memory functions (MQ = 49). Difficulties with auditory sequential memory and reproduction of spatial designs from memory were apparent. Spatial memory for designs was within normal limits in the right hemidecorticate, although verbal memory was impaired (MQ = 67). Since posthemispherectomy IQs in both of these cases were reduced by nearly half compared with preoperative estimates, Gott (1973b) concluded that the combined operation of the two hemispheres facilitates the development of new functions in the developing brain.

Day and Ulatowska (1979) reported an impairment in auditory sequential memory in a right hemidecorticate while visual sequential memory approached the normal range. The authors did not offer an explanation for this unexpected disparity. Similarly, the left hemidecorticate exhibited a severe auditory memory span deficit, while visual memory was less impaired.

1.3.4d. Intellectual Level. Table 1 summarizes the data from studies reporting Wechsler Verbal and Performance IQs. It is seen that the infant and child groups exhibit comparably depressed scores independent of the side hemispherectomized. This pattern is similar to the nonspecific intellectual deficit

Table 1. Posthemispherectomy Verbal and Performance IQ Scores Averaged across Published Series of Patients[a]

Age at onset of brain damage		Left hemispherectomy			Right hemispherectomy	
		Verbal IQ	Performance IQ		Verbal IQ	Performance IQ
Infant	(*N* = 14)	86.5	81.4	(*N* = 11)	79.5	78.5
		(16.7)	(16.1)		(14.2)	(22.2)
Child	(*N* = 6)	72.8	69.0	(*N* = 5)	89.8	71.4
		(20.9)	(19.7)		(19.3)	(14.2)
Adult	(*N* = 2)	—	102.5	(*N* = 5)	102.6	69.6
		—	(7.8)		(12.9)	(5.0)

[a]The IQ scores are mean values obtained by averaging the data of patients in the same age group. Standard deviations are given in parentheses. Severe expressive aphasia in adults with left hemispherectomy precluded administration of the verbal scale of the Wechsler Intelligence Test.

which persists after the onset of infantile hemiplegia (Woods, 1980). In contrast, adult right hemispherectomies exhibit marked visuospatial deficit with well-preserved verbal ability and left hemispherectomy produces a severe aphasia. As noted by Gott (1973b) and Smith (1974) and depicted in Table 1, lateralized IQ differences do occur in adult cases, including patients with prolonged recovery periods similar to that typically found in children.

McFie (1961) examined cognitive outcome in 34 infantile hemiplegics assessed before and after hemispherectomy. Statistically significant increases in IQ were observed only in children sustaining injuries in the first year of life. Changes in IQ were not associated with either preoperative IQ or age at operation. Both left and right hemidecorticates, particularly those sustaining brain damage in the first year of life, scored higher on performance than verbal subtests. This finding is atypical since most studies report higher verbal IQs (e.g., Griffith and Davidson, 1966; Carlson *et al.*, 1968; Smith and Sugar, 1975). McFie's (1961) suggestion of preferential recovery following infantile injury is negated by methodological problems. No controls were included to estimate test–retest variability, which is a serious omission in view of the small sample size. The Wechsler-Bellevue scale was administered to only 18% of the cases since the remainder exhibited marked intellectual impairment. Based on EEG recordings, only 38% of the sample had normal remaining hemispheres which suggests bilateral involvement and is consistent with the low mean IQ of 60 in this sample.

1.3.4e. Motor Functions. It has been found that approximately 91% of the hemispherectomized children achieve independent ambulation. No useful movement of the affected upper extremity was reported in 11 of 21 pediatric cases while the remaining ten could perform gross tasks and had limited ability to open the hand and fingers (Wilson, 1970; Damasio et al., 1975; Ignelzi and Bucy, 1968; Gott, 1973a,b).

1.3.5. Outcome of Hemispherectomy in Adults

Burkland and Smith (1977) and Smith (1974) described two left-hemispherectomized adults. Receptive language skills were least affected and showed the greatest improvement over time. Expressive skills were severely impaired although one case (EC) was able to speak single words immediately after operation. Continuing improvement was noted in receptive and expressive functions until EC was able to speak in short propositional sentences. Production and comprehension of written language remained severely impaired, precluding administration of the verbal scale of the Wechsler Intelligence Test. Since tumor regrowth occurred in both cases within approximately two years, it is unknown to what extent language functions might have developed. In both cases, impairment of nonverbal functions was also evident.

Early psychometric studies of right-hemispherectomized adults were misleading regarding cognitive outcome since patients were assessed only on language-based instruments such as the Stanford-Binet (Smith, 1966). Assessment of nonlanguage functions revealed persistent impairment in the integration of visual information (Smith, 1974) and nonverbal reasoning (Gott, 1973b; Bruell and Albee, 1962; Smith, 1974). In contrast to persistent nonverbal deficits, language (Bruell and Albee, 1962; Gott, 1973b; Mensh et al., 1952) and speech functions (Bell and Karnosh, 1949) were retained. Verbal reasoning and other verbally mediated skills such as vocabulary and arithmetic functions were not altered (Smith, 1974).

Deficits in visual retention occur independently of laterality of hemispherectomy (Smith, 1974; Burkland and Smith, 1977). Bell and Karnosh (1949) reported no change in memory functions although formal assessment was not undertaken, while Bruell and Albee (1962) identified deficits in remote memory. Four right-hemispherectomy cases have been tested on the WMS which primarily examines short-term verbal memory. The mean memory quotient (MQ) was 88.5 and the range extended from 62 to 126 (Gardner, 1955; Mensh et al., 1952; Gott, 1973b). Since IQ data were not reported for most cases, comparison of memory functioning with verbal and performance IQs is not possible. However, Gott (1973b) reported a MQ of 77, which is lower than would be expected given a verbal IQ of 105.

Independent ambulation was reported in nine of ten adult cases reviewed (Gardner, 1955; Bruell and Albee, 1962; Burklund and Smith, 1977; Smith, 1966; Gott, 1973b; Bell and Karnosh, 1949). Functionally useful movement of the affected extremity is uncommon (Mensh et al., 1952). While some independent movements of the arm and fingers have been reported, they are not associated with a useful grasp (Smith, 1966; Burkland and Smith, 1977; Bell and Karnosh, 1949).

In summary, hemispherectomy in adults is associated with comparatively lateralized cognitive disturbances since previously acquired functions of the intact hemisphere are minimally affected. In contrast, removal of either immature hemisphere after damage in infancy results in less specific deficits. Although the immature brain compensates to some degree for these deficits, a generalized underdevelopment of cognitive abilities is frequently present.

At the present time it is not possible to make inferences about the relationship between age and memory dysfunction since the data are limited. Memory deficits do not appear to be related to the side of hemispherectomy since verbal deficits result from removal of either hemisphere in childhood. In adults, visual retention is poor independently of the side hemispherectomized, and verbal deficits are observed in some right-hemispherectomy cases. However, motor deficits of both the upper and lower extremities are severe independently of age at hemispherectomy. Consistent with the animal literature, these findings suggest that the degree of maturation of the lesioned substrate is related to the severity of deficits. Motor functions are comparably affected independently of age at lesion or hemispherectomy. Language functions, which develop late ontogenetically, are least affected by early injury, suggesting that both anatomical and functional reorganization are maximized in later developing functions.

1.4. Diffuse Insult to the Young Brain

1.4.1. Infectious Disease

The generality of functional plasticity must be qualified to accommodate the results of studies of diffuse insult to the young brain due to encephalitis, meningitis, and Reye's syndrome (Davidson et al., 1978; Wright, 1978). Residual cognitive deficit secondary to infection is more severe when the onset of illness is during early infancy as compared with later childhood. Neurologic and intellectual sequelae are particularly common after meningitis occurring during the first year of life. Wright (1978) concluded from an extensive review of the literature that "the younger the victim, the greater the impact of meningitis on his/her intellectual functioning" (p. 1038). Consistent with these findings, Davidson and colleagues (1978) found that the age of onset of Reye's syndrome was also related to outcome. All children who recovered with average to above average IQ were at least two years old at the time of onset, leading the authors to infer that age was the most useful prognostic index. In a study of outcome from illnesses other than head injury which had produced coma in children, Seshia, Seshia, and Sachdeva (1977) reported that seizures occurred more frequently in the 13- to 36-month age range, but that age was not related to global outcome. The authors, however, did not include cognitive assessment in their study. On balance, there is no evidence that young children recover more fully from diffuse, infectious insult to the brain than adults.

1.4.2. Vegetative State after Hypoxia–Ischemia

The term persistent vegetative state was used by Jennett and Plum (1972) to describe a clinical syndrome after brain damage in which patients seem to be in a state of "wakefulness without awareness," showing no evidence of meaningful interaction with the environment. Gillies and Seshia (1980) serially studied 17 children under three years of age who had become vegetative after a period of coma precipitated by a hypoxic–ischemic illness or meningitis. The condition of vegetative state carried a poor prognosis in that seven children died of complications (e.g., aspiration) and the survivors displayed no cognitive functioning a year after the onset of illness. The overall outcome of these children was no better than that reported for persistently vegetative head-injured adults.

1.4.3. Cranial Irradiation

Concern about the effects of whole brain irradiation (a treatment which potentially has widespread impact on cerebral functioning) in children with acute lymphocytic leukemia has led to several studies of cognitive ability in survivors (Eiser, 1978; Ivnik et al., 1981; Meadows et al., 1981). While the treatment regimen in addition to cranial irradiation has not been consistently controlled in all studies, the evidence points to impairment in reasoning, arithmetic, short-term memory, visuospatial ability, and visuomotor speed in children given prophylactic cranial irradiation as compared with children who have delayed cranial irradiation or those who do not receive this treatment. These studies have shown that the degree of cognitive impairment is most impressive in children who were two to five years old when cranial irradiation was administered as compared with older children receiving similar treatment (Meadows et al., 1981). Consistent with the evidence for long-term deleterious effects of cranial irradiation on cerebral functioning, follow-up study has demonstrated persistent abnormalities of the visual-evoked response (Yaar et al., 1979). The pattern of findings parallels the effects of age on recovery from diffuse infectious insult to the brain.

1.4.4. Malnutrition

Similar vulnerability of the immature brain has been suggested with respect to the effects of malnutrition. Winick (1976) concluded from his review of this topic that the most adverse effects are present in malnourished children who are younger than two years.

1.4.5. Epilepsy

The age of onset of convulsive seizures is a major determinant of cognitive functioning in epileptics. However, in many studies it is difficult to differentiate the effects of age of onset from the duration of epilepsy and the number of seizures. Dikmen, Mathews, and Harley (1975) gave neuropsychological tests to adult patients with motor seizures and reported that those with a late onset (10–25 years old) had intellectual functioning above the level of patients with onset before their sixth birthday. The adverse effects of early onset of tonic–clonic seizures were also demonstrated in children by O'Leary *et al.* (1981). The authors observed that onset before age five led to more severe cognitive impairment on tests of problem solving, complex sensorimotor performance, and short-term memory as compared with children with onset of seizures after age five. In a study of the long-term outcome of childhood temporal lobe epilepsy, Lindsay, Ounsted, and Richards (1979) found that when seizures began before four years of age they persisted into adulthood, while remission was more common in children whose seizures began later in life. In summary, these studies indicate an inverse relationship between the age of onset of epilepsy and subsequent cognitive functioning.

2. HEAD INJURY THROUGH THE LIFESPAN

2.1. Incidence

The two main measures of the frequency of a disorder in the community are incidence and prevalence. The incidence of head injury is the number of new cases occurring in a defined population within a designated time period (usually a year). Incidence is expressed as the number of new cases relative to the individuals at risk for developing such an injury. This occurrence rate is the probability of a person at risk sustaining a head injury during the specified time period. The prevalence of head injury is the frequency of existing cases (including the number of new patients) in the defined population who survived a previous head injury during a designated time period.

The published epidemiologic data on head injury occurring at various ages are compromised by omission of patients with mild head injury who were treated in the emergency room and fatal injuries who died before they could be evacuated to a hospital (Kraus, 1980). Few studies of incidence have applied specific criteria of head injury severity to a total population in whom public health records (e.g., death certificates) were available in addition to hospital data.

The first published study of incidence which applied clinical criteria of head injury severity to a defined population is that of Annegers, Grabow, Kurland, and Laws (1980) for Olmsted County, Minnesota. The authors used an extensive

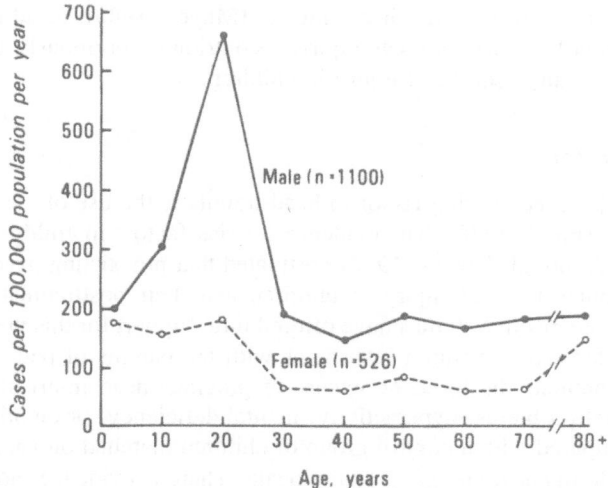

Figure 2. Incidence rates of head injury in Olmsted county, Minnesota, 1965–1974. [From Annegers *et al.* (1980). Reproduced with permission of the authors and publisher.]

countywide medical records system to identify patients with head injury who had evidence of "presumed brain involvement," i.e., loss of consciousness, posttraumatic amnesia, neurologic signs of brain injury, or skull fracture. The overall incidence per 100,000 was 274 in males and 116 in females. Figure 2, which shows the age and sex-specific incidence rates for head injury in Olmsted County, Minnesota, indicates a steep rise at age 15–24 (particularly for males), followed by a progressive decline until a secondary peak occurs after age 70. While overall head injury admissions in children reach a peak at ages 4–8 (Craft, 1972), Moyes (1980) found that admissions for severe head injury had peaks at ages three, eight, and ten. As shown in Fig. 2, a later peak in head injury admissions occurs after age 70.

Age is also related to the cause of head injury. Young adults and late adolescents are most often injured as occupants of motor vehicles. In contrast, young children are prone to pedestrian–car accidents and to falls.

2.2. Cause of Head Injury

Hospital-based studies of head injury in adolescents and adults show that road traffic accidents account for about one half of all closed-head injury (CHI) cases (Annegers *et al.*, 1980) and that this etiology is strongly implicated in injuries producing coma. Falls contribute substantially to head injury in children (nearly half of the cases studied by Rowbotham *et al.*, 1954) but account for less than a third of severe pediatric head injuries (Moyes, 1980). Road traffic accidents, particularly pedestrian–car injuries, contribute enormously to the mortality and morbidity after head injury in children.

2.3. Risk Factors

The major predisposing factor in head trauma is the use of alcohol shortly before injury (Field, 1976). The evidence for risk factors in children is far less compelling. Although Jennett (1972) postulated that preexisting psychiatric disorder contributes to head injury in children and their posttraumatic behavior disturbance, prospective studies have offered little support for this view. Klonoff (1971) reported that structured interviews with the parents of pediatric closed-head injury patients disclosed no excess of previous head injuries, antecedent developmental problems, hyperactivity, mental deficiency, or emotional disturbance as compared with a control group of children matched on age, sex, grade in school, intelligence, and geographic locale. There is evidence, however, that the incidence of subsequent head trauma after an initial head injury is above expectation for the patient's age and sex (Annegers *et al.*, 1980). The authors postulated that a behavioral pattern predisposing to injury may develop. In this connection, it is conceivable that residual deficits (e.g., slowed reaction time) may increase the vulnerability to head injury.

3. PATHOPHYSIOLOGY OF CLOSED-HEAD INJURY IN RELATION TO AGE

The reportedly greater capacity of young children to survive severe CHI as compared with adults may be attributed to a number of anatomic and physiologic features of head injury in the pediatric age group (Bruce *et al.*, 1978; Hendrick *et al.*, 1964). The greater flexibility of bones in young children may enhance the capacity of the skull to absorb traumatic forces, thereby reducing focal brain injury (Craft, 1972; Gurdjian and Webster, 1958).

There is growing evidence that children exhibit more severe acute neurologic dysfunction. Gurdjian postulated that the relatively shallow cerebral convolutions of the child's brain would result in greater deformation of the brain on impact, potentiate shearing effects, and contribute to brain stem injury. Accordingly, Bruce, Schut, Bruno, Wood, and Sutton (1978) found that oculovestibular function was impaired or absent in 30% of pediatric CHI cases who fulfilled the criteria for severe injury according to the Glasgow Scale, i.e., inability to obey commands, utter comprehensible speech, or to open eyes for a period of at least six hours after injury. Nearly one third exhibited bilaterally fixed pupils at the time of admission. In their series, fewer mass lesions (e.g., intracerebral hematomas) occurred than in adults with injuries of comparable severity (Becker *et al.*, 1977; Jennett *et al.*, 1977). However, diffuse cerebral swelling with oblitera- tion of the ventricles and cisterns (Fig. 3, left) followed by enlargement of the ventricles to normal size (or larger) over time (Fig. 3, right) was found in one third of the cases. In view of the low mortality in this series, the severity of the neurological deficits was remarkable. These findings led Bruce and colleagues (1978) to infer that CHI in children results primarily in diffuse injury to the white matter. They suggested that cerebral swelling reflects increased cerebral blood volume which ultimately impairs intracranial compliance to changes in pressure.

In view of the grave acute neurologic condition and relatively favorable prognosis for survival in their series, Bruce and his co-workers (1978) concluded that "If the threshold for neurophysiological dysfunction is lower in children than in adults, then for the same input force to the brain, the recorded neurologi- cal picture will be worse in the child." An implication of this view is that head injuries in children and adults may not be strictly comparable even if the initial Glasgow Coma Scale scores are equal. As mentioned earlier, the etiology of injury is often different in young children as compared with adults. The hetero- geneity in mechanisms of injury warrants consideration in studies comparing the outcome of different age groups.

Children exhibit acute manifestations of mild CHI which also differ from those found in adults. Hendrick *et al.* (1964) characterized the acute mental status of children with mild CHI as a condition of drowsiness, confusion, and irritability. Anterograde and retrograde amnesia were reported by Klonoff and Low (1974) to be less frequent in pediatric CHI patients under nine years old than in older children. However, this apparent age-related disparity may reflect ambi- guities in evaluating the presence of amnesia in young children.

4. EFFECTS OF AGE ON OUTCOME AFTER HEAD INJURY

The studies of outcome after severe CHI have shown a lower mortality in children as compared with adults. Figure 4, which is based on the findings of an international data bank (Teasdale *et al.*, 1979), depicts a rise in mortality as the age of the patient increases. Consistent with this pattern, Bruce and colleagues (1978) found a 6% mortality in a series of 53 children admitted to the Children's Hospital of Philadelphia after sustaining severe head injury (Glasgow Coma Scale score of ≤8 on admission). Although the likelihood of an intracranial hematoma complicating severe head injury increases with age, this factor alone does not account for the large differences in mortality across the age range (Teasdale *et al.*, 1979).

The effects of age on the quality of recovery after CHI are less consistent across studies than are the mortality data. In general, the clinical picture of long-

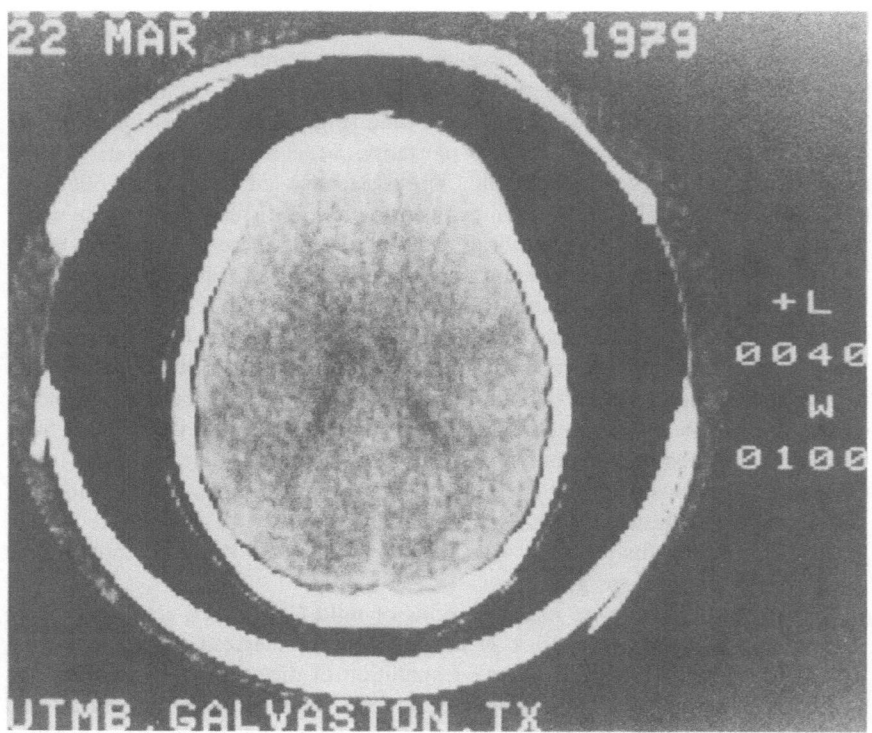

Figure 3. Computerized tomographic scans obtained on the day of a severe head injury (22 March 1979) (left) and two months later (9 May 1979) (right) in an eight-year-old boy. The initial Glasgow Coma Scale score was 7; he was unable to obey simple commands for 40 days postinjury. Initial scan

term recovery appears to be less favorable when assessment of cognitive function and measures of school achievement are included in the evaluation of outcome.

Bruce and colleagues (1978) reported that 90% of a series of pediatric CHI patients showed a good recovery or only moderate disability after injury which produced coma (according to the Glasgow Scale) and pervasive neurologic deficit. Although Bruce and co-workers focused on the pathophysiology and clinical management of acute CHI in children, they conveyed the impression of relatively minor residual deficit after severe injury. A contrasting impression emerges from the study by Brink, Imbus, and Woo-Sam (1980) who reported data on physical and cognitive recovery one year after severe head injury (coma of at least 24 hours) in 344 children and adolescents who were attending a rehabilitation program. The authors noted that less than 10% of the series had normal neurologic findings, and that cognitive function remained impaired in about two thirds of

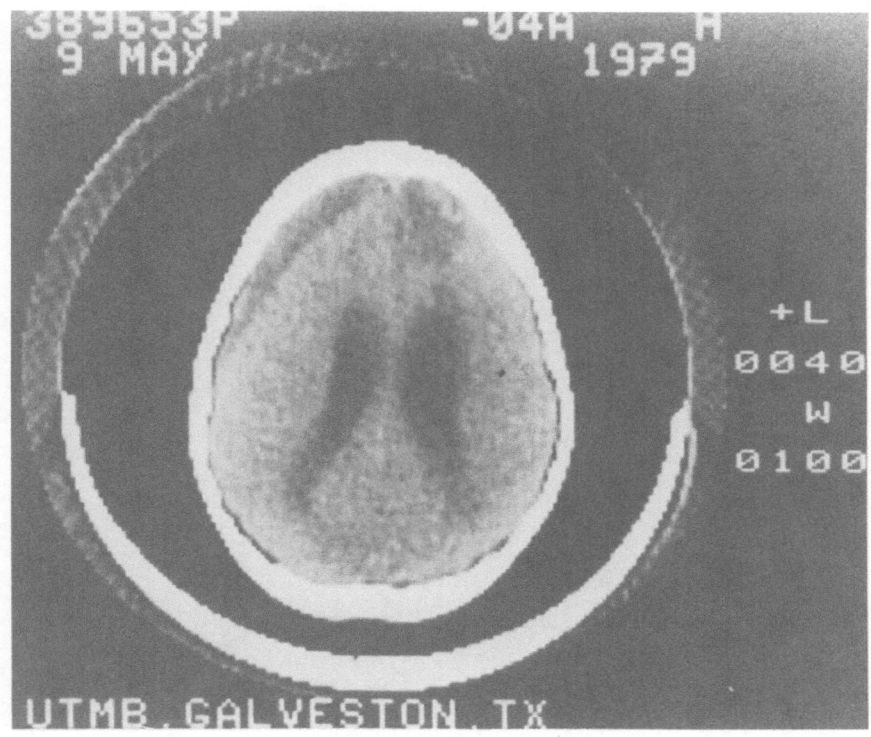

shows initial cerebral swelling, reflected by the obliterated lateral ventricles that subsequently enlarged. [From Levin *et al.* (1982). Reproduced with permission of the publisher.]

those cases. In contrast, nearly three fourths of the total sample recovered ambulation and were capable of self-care.

    This disparity in outcome findings suggests that observations of improved
motor function and adjustment to activities of daily living in children who have
sustained severe head injury may lead one to overestimate the quality of cognitive recovery. As will become clear in the literature review, cognitive impairment frequently persists after severe injury despite the disappearance of focal
motor and sensory deficits and the resumption of daily activities. Behavioral
changes after CHI may be classified according to their time course. "Early"
effects appear after termination of coma, whereas "late" effects are manifested
after resolution of confusion and may involve relatively permanent changes.
Although previous studies have suggested that grossly aberrant behavior is frequently transient and confined to the initial phase of recovery, the contention that
"significant sequelae" are rare is unsupported by the evidence.

n: 44 46 94 63 57 46 39 49 43 43 38 39 32 16 19

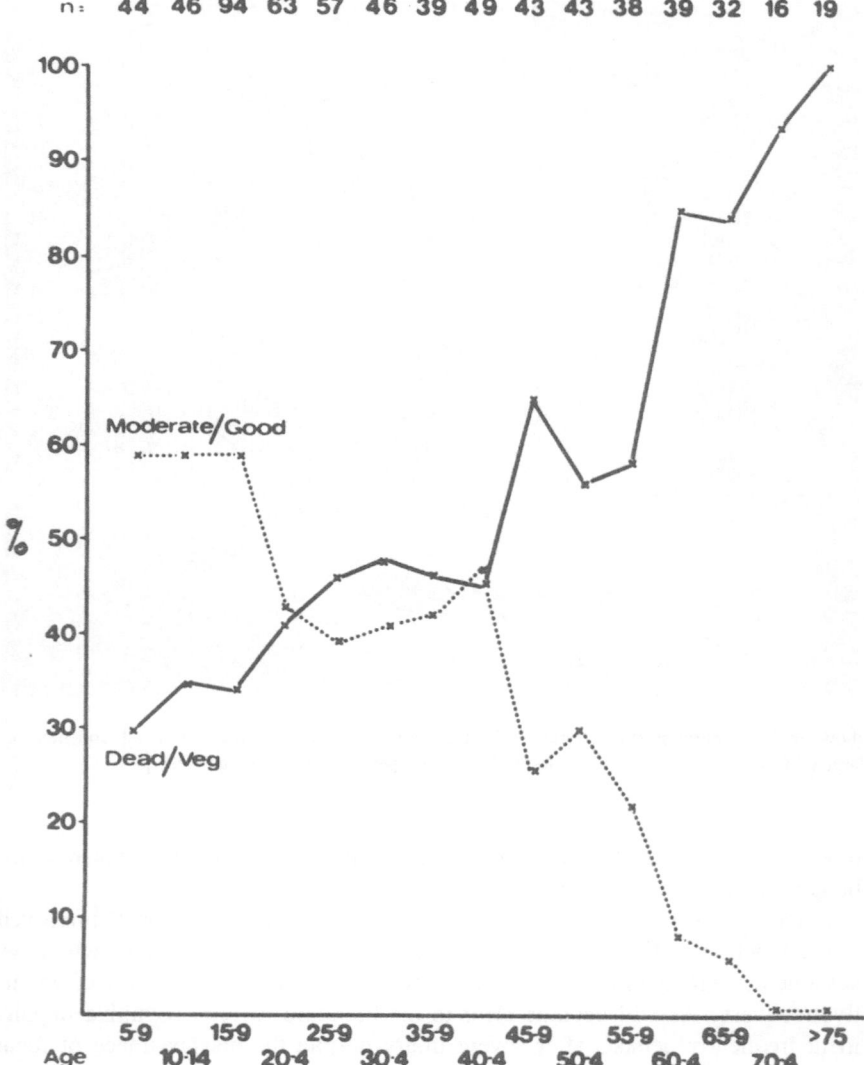

Figure 4. Outcome of severe head injury plotted against age at the time of injury. Based on cases from the International Coma Data Bank. The percentage of patients with a good recovery or moderate disability declined as the age of injury increased, while the percentage of patients who died or remained persistently vegetative after a severe head injury increased as a function of age at injury. [From Jennett and Teasdale (1981). Reproduced with permission of the authors and publisher.]

5. EARLY AND LATE NEUROBEHAVIORAL EFFECTS OF HEAD INJURY

5.1. Subacute Behavioral Effects after Head Injury in Children

During the initial phase of recovery, children show signs of confusion, disorientation, and possibly paramnesia, (e.g., misidentification of geographic location) similar to the manifestations described after head injury in adults. The transition from coma to normal orientation may be accompanied by anterograde and retrograde amnesia, lethargy, and akinesia (Klonoff and Low, 1974; Todorow, 1975). There are no standardized procedures to assess amnesia in young children; age-based norms for temporal orientation are unavailable. Heightened irritability, drowsiness, and confusion may dominate the mental status in children who maintain consciousness (Hendrick et al., 1964). Todorow described an "akinetic mutistic state" in which the child copes with feelings of helplessness and restraints imposed by life-support equipment by withdrawing from the environment. The author interpreted the child's effort to remain motionless and apathetic despite signs of neurologic improvement as an attempt to escape a frightening situation.

Blau (1936) described six children with posttraumatic psychosis who required transfer from the neurosurgical service to a psychiatric unit for periods of two to four weeks. The manifestations included acute excitement with impulsiveness and restlessness. The children were noisy, attempted to get out of bed, and cried or screamed continuously. The outstanding feature of these cases was a period of unrestrained emotional and motor behavior accompanied by severe anxiety. Posttraumatic amnesia was present in all six children, but they were able to recognize members of their family shortly after regaining consciousness. When the children were examined at follow-up at least six months postinjury, the posttraumatic psychosis had resolved.

5.2. Residual Neuropsychological Sequelae

In view of the high incidence of CHI among children, there have been relatively few investigations of neuropsychological recovery. The studies published to date are compromised by nonuniform criteria for grading the neurologic severity of injury and defining coma; "severe" CHI in children has been inferred from a duration of coma ranging from at least 30 minutes (Klonoff and Low, 1974; Klonoff, Clark, and Low, 1977) to at least one week (Brink et al., 1970; Chadwick et al., 1981). Richardson (1963) defined severe injury on the basis of a posttraumatic amnesia (PTA) duration which exceeded seven days. The Glasgow Coma Scale has seldom been used in neuropsychological studies to document acute neurologic impairment (Levin and Eisenberg, 1979a,b). Methodological problems have also arisen from the wide range of assessment techniques. Several studies are limited to the reporting of overall intellectual level (cf., Brink *et al.*, 1970). Specific testing of learning and memory has been less often performed (Fuld and Fisher, 1977; Levin and Eisenberg, 1979a,b; Levin *et al.*, 1982). In view of the claim that children recover after CHI more readily than adults, the time course of study becomes a major consideration. Consequently, serial testing of children (Klonoff and Low, 1974) or at least evaluation after an extended interval (Brink *et al.*, 1970; Richardson, 1963) are essential to evaluate the long range implications of findings during earlier stages of recovery. Our review of the neuropsychological effects of CHI is organized according to the function investigated.

5.2.1. Language

Lenneberg (1967) concluded from his review of the clinical literature that "the chances for recovery from acquired aphasia are very different for children than for adult patients, the prognosis being directly related to the age at which the insult to the brain is incurred" (p. 142). He postulated that language acquisition is dependent upon cerebral maturation and that the potential for recovery from aphasia deteriorates rapidly after age 12. While quantitative studies of recovery from acquired aphasia in head injured children have generally supported Lenneberg's postulation, the data indicate that qualifications are necessary.

Hécaen (1976) studied 26 cases of acquired aphasia in children, including 16 cases of head trauma who ranged in age from six to 16 years and who were studied over varying periods. Although the series included four children considered to have primary right-hemisphere damage, bilateral injury could not be excluded in these cases. Aphasic disturbance was characterized by a period of mutism which persisted for as long as three months, reduced initiation of speech, and the absence of paraphasias. Anomia was frequently present during the acute period and tended to persist, whereas impairment of auditory comprehension was uncommon and recovered rapidly. Alexia also resolved within a brief period, but disturbance of writing was more prolonged. Acalculia was a frequent concomitant. Hécaen concluded that acquired aphasia has a better prognosis in children than in adults.

Levin and Eisenberg (1979b) evaluated language deficit after CHI by administering the Neurosensory Center Comprehensive Examination for Aphasia (Spreen and Benton, 1969) to 64 children and adolescents following resolution of posttraumatic confusion. Language defects were present during the first six months after injury in nearly one-third of the series. Similar to the pattern of aphasic defects found in adults after CHI, anomia was the most common deficit and impairment of repetition was rare. Comprehension of oral language was impaired in only 11% of the patients. Follow-up examinations showed impressive recovery in young children, whereas residual deficits persisted when patients were injured during late adolescence. Closed-head injury associated with focal left-hemisphere injury (e.g., hematoma) was compatible with either full recovery or relatively isolated residual defects such as anomia or deficient word retrieval.

Recent findings suggest that subtle language defects may persist after severe head injury in children. Chadwick and colleagues (1981) reported that object-naming latency was slowed a year after head injury which produced a period of PTA of at least a week. Woods and Carey (1979) found that children with left-hemisphere brain damage who had apparently recovered from aphasia exhibited residual deficit on tests of picture naming, spelling, sentence completion, the usage of verbs, complex comprehension (Token Test), and syntax. The age at injury was an important determinant in language deficit insofar as children who sustained a left-hemisphere insult after their first birthday had more pervasive language disturbance. In contrast, children with early injury were impaired only on the spelling test.

After the first birthday has passed, however, the effects of age on recovery from aphasia may be relatively minor in comparison with the etiology of brain damage and the type of aphasia. In a follow-up study of acquired aphasia in children, van Dongen and Loonen (1977) found that complete recovery was common after head injury, while patients with vascular disease remained aphasic. The type of aphasia was also contributory in that children with a mixed expressive–receptive picture fared worse than children whose initial language disturbance was confined to dysfluent and anomic disorders. Residual impairment of oral expression, reflected by a reduction in spontaneous speech, was reported by Gaidolfi and Vignolo (1980) in four of 21 patients when they were examined nearly ten years after they had sustained a CHI which produced coma. On balance, studies of acquired aphasia in head-injured children confirm Lenneberg's assertion that aphasia in children is less permanent than aphasia occurring in adulthood. On the other hand, quantitative assessment of language has shown that residual defects are common after apparent restoration of language in children.

5.2.2. Memory

Few studies have assessed the capacity of head-injured children to store and retrieve new information. Richardson (1963) tested ten CHI patients (five to 18 years old) with "very severe concussion" who had been in coma more than seven days and were examined at least 18 months after injury. Results based on selected items of the Stanford-Binet and the Benton Visual Retention Tests disclosed memory deficit in all of the patients despite improvement in fine motor skills and in adjustment to daily activities. The importance of assessing long-term storage and retrieval was also shown by Fuld and Fisher (1977) who serially studied two pediatric CHI patients and administered the selective-reminding test (Buschke and Fuld, 1974). Consistent with Richardson's results, the authors found that the recovery of long-term memory lagged behind the resolution of motor deficit.

In a series of studies completed in Galveston, Levin and colleagues (Levin and Eisenberg 1979a,b; Levin and Grossman, 1976; Levin *et al.*, 1982) employed a modified version of the selective reminding test (Fig. 5) and a continuous recognition memory test to evaluate residual memory deficit in children and

| NAME: Control ♀ | | | | DATE: 2-79 | | | |
| AGE: 9 | | | | EDUC: 3 | | | |

SELECTIVE REMINDING FOR
CHILDREN ≤ 12 y.o.

	1	2	3	4	5	6	7	8
Dog		1→2	2	3	2	2	2
Fox	1→6	1	1	2	1	1	1
Horse		5→10	9	10	3	3	3
Lion		2→8	10	1	12	8	12
Elephant	3→10	6	3	8	11	7	7
Bear	4→12	9	6	7	9	9	4
Rat	2	11		5→9	10	10	9
Raccoon		3→3	8	11	8	12	6
Goat		4		4		6→11	8
Squirrel	7	8	7		4→5	5	10
Beaver	5	9	4		5→4	6	5
Turtle	6→7	5	7	6	7	4	11
Total Recall	7	12	10	10	11	12	12	12
LTR	7	11	10	9	11	12	12	12
STR	0	1	0	1	0	0	0	0
LTS	7	11	11	11	11	12	12	12
CLTR	4	8	8	9	11	12	12	12
Random LTR	3	3	2	0	0	0	0	0
Presentations	12	5	0	2	2	1	0	0

Figure 5. Modified version of Buschke's Selective Reminding Test for children. The heavy underline denotes the long-term storage of a word, whereas the arrow indicates the beginning of consistent retrieval from long-term storage, i.e., the word is retrieved on all successive trials without further reminding by the examiner. LTR, long-term retrieval; STR, short-term retrieval; LTS, long-term storage; CLTR, consistent long-term retrieval.

adolescents after clearing of PTA. As depicted in Fig. 3 of Levin and Eisenberg (1979a), long-term storage and retrieval, as measured by the selective-reminding test, were particularly affected by CHI; nearly one half of the patients with CHI of varying severity had an impairment of memory, which was inferred from scores that fell below the fourth percentile of control subjects of comparable age.

The authors reported that CHI which produced coma (unresponsiveness to commands) for periods of 24 hours or more resulted in greater impairment of consistent retrieval from long-term storage than mild injury which caused only momentary or no loss of consciousness and no neurologic deficit. Levin and Eisenberg (1979a,b) also found that the proportion of patients with residual memory deficit increased according to the severity of acute injury as measured by the Glasgow Coma Scale. An excessively high number of false-positive errors characterized the continuous recognition memory of severely injured children and adolescents, a pattern similar to that found in young adults after CHI (Hannay et al., 1979).

Consistent with the Galveston studies, Gaidolfi and Vignolo (1980) reported evidence of a verbal memory deficit nearly ten years after injury in nearly one quarter of children who had sustained a CHI which produced coma. Memory span for spatial location was less frequently impaired.

5.2.3. Comparison of Recovery of Memory in Children and Adolescents

To investigate the relative recovery of memory in young children (median age = 8) and adolescents (median age = 17) after severe CHI, Levin and colleagues (1982) individually matched patients from these age groups on the type of injury (mass lesions, diffuse injury) and the initial Glasgow Coma Scale score. Figure 6 shows that the long-term recovery of verbal memory on the selective-reminding test was less impressive in children as compared with adolescents, whereas baseline performance was markedly impaired in both age groups. This disparity in long-term recovery was also found for continuous recognition memory (Fig. 7). It is seen that follow-up assessment at least six months postinjury disclosed that the performance by severely injured adolescents approximated the recognition memory of teenagers who sustained mild to moderate injuries (Glasgow Coma Scale score > 8). These findings suggest that the restoration of memory after diffuse insult to the young brain is slower and less complete than the recovery of language after focal left-hemisphere injury at this age.

5.2.4. Intellectual Functioning

Marked decline in scholastic achievement and transfer to a special education program have been documented in follow-up studies of children with severe head

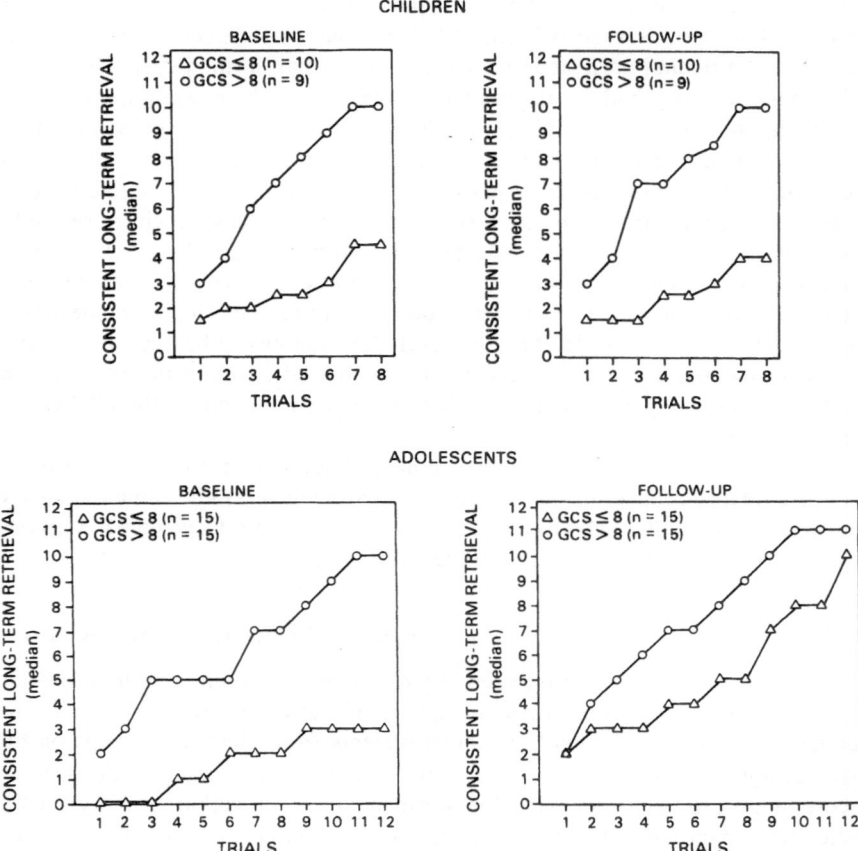

Figure 6.   Baseline and follow-up verbal learning and memory on the Selective Reminding Test given to head-injured children and adolescents. Patients sustaining severe head injury (Glasgow Coma Scale score ≤8) exhibited impaired consistent retrieval from long-term storage in comparison with patients of similar age who sustained less severe head injuries. The follow-up testing was done at least six months postinjury. [From Levin *et al.* (1982). Reproduced with permission of the authors and publisher.]

injury (Brink *et al.*, 1970; Fuld and Fisher, 1977; Heiskanen and Kaste, 1974; Klonoff, Low, and Clark, 1977). Deterioration in school performance was related to neurologic deficits such as ataxia and aphasia, as well as to neuropsychological sequelae (Brink *et al.*, 1970; Klonoff *et al.*, 1977; Klonoff and Low, 1974). In a longitudinal study of 231 children, the majority of whom sustained mild CHI, Klonoff and Clark (1977) found progressive increments in IQ over a five-year period in both children younger and older than nine years at

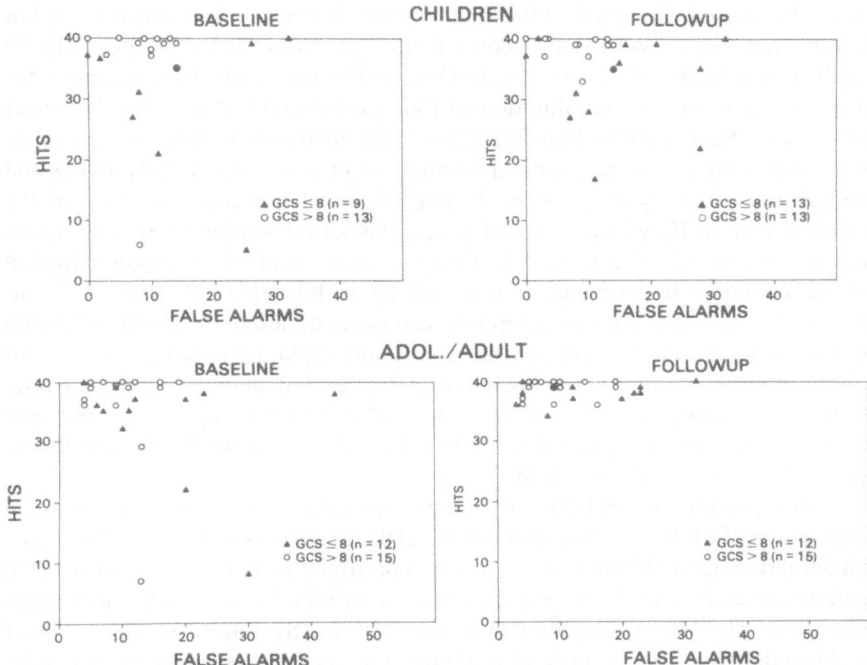

Figure 7.   Visual recognition memory performance on baseline and follow-up testing. Hits (correct identification of recurring pictures) are plotted against false alarms (misidentification of a new picture as one previously seen) for children and adolescents with head injury. It is seen that severe head injury (Glasgow Coma Scale score ≤8) resulted in a marked rise in false alarm errors with relatively few hits as compared with milder head injury. There is also a trend of greater improvement after severe head injury in adolescents as compared with children. [From Levin *et al.* (1982). Reproduced with permission from the authors and publisher.]

the time of the injury. In an earlier study, Klonoff and Low (1974) reported an average decrement of 10 IQ points when evaluation was performed one year after injury. Chadwick and colleagues (1981) confirmed the presence of intellectual deficit on the performance scale of the Wechsler Intelligence Test given one year after severe head injury.

Consistent with parallel studies of severely head-injured adults, marked cognitive impairment has been reported in follow-up studies of children who had sustained severe CHI. Flach and Malmros (1972) found that nearly one half of 131 patients exhibited cognitive deficit when tested eight to ten years after injury. In conformity with the one-year follow-up data of Chadwick *et al.*, the authors reported that the performance IQ was most affected. In a ten-year follow-up study of 21 pediatric CHI patients who had been in coma for varying durations, Gaidolfi and Vignolo (1980) found that four cases had Wechsler Adult Intelligence Scale (WAIS) IQ scores below 80. Brink and colleagues (1970) studied 52 patients (age range, two to 18 years at injury), who were comatose for at least one week (median duration of four weeks) and tested on the Wechsler Intelligence Scale for Children (WISC) or Stanford-Binet one to seven years after injury. In contrast to the eventual resumption of activities of daily living and ambulation in nearly all patients, testing disclosed that only one third of the children had an IQ which was within one standard deviation of the population mean (i.e., an IQ of at least 85). Thirty-seven percent of the group exhibited severe cognitive impairment, as reflected by an IQ below 70. Brink and colleagues found a direct relationship between coma duration and residual IQ with greater impairment in children who were under eight years old at the time of injury. Further evidence for unequivocal intellectual deficit in children after severe head injury has been obtained in studies which compared the outcome measures with preinjury test data provided in school records (Levin and Eisenberg, 1979b; Richardson, 1963).

The greater vulnerability of infants and toddlers to long-term cognitive impairment (four to 14 years after severe CHI), as compared with preschool- and kindergarten-aged children, was recently confirmed in a follow-up study of 50 patients treated on the neurosurgical service at the Klinikum, West Berlin (Lange-Cosack *et al.*, 1979). Duration of coma was directly related to the extent of residual deficits. The presence of acute brain injury was frequently visualized by CT findings, and serial CT scans disclosed evidence of ventricular dilatation in several cases. In a similar study, Levin and colleagues (1982) showed that severe head injury (Glasgow Coma Scale score < 8) frequently results in subnormal intellectual level (IQ < 80) in children under 12 years old at the time of injury, whereas the residual intellectual level was higher in adolescents with comparable Glasgow Coma Scale scores. In view of the agreement between the findings reported by Brink and her colleagues and the Texas studies, there appears to be an inverse relationship between age and intellectual function after severe head injury. Brink, Garrett, Hale, Woo-Sam, and Nickel (1970), who administered the WISC to ten severely head injured children (minimum PTA of seven days, median coma duration of 28 days) at intervals ranging from 1.5 to 13.5 years after injury, found evidence of intellectual deficit in two thirds of the series. Richardson (1963) found that the IQ of the patients ranged from the borderline-defective level to the low-average range. Preinjury data indicating that nine of the patients had previously functioned at the average level or better led the authors to estimate that there had been a decrement of 10 to 30 points in IQ.

In summary, the results offer no support for the view that children are spared residual cognitive deficit after severe CHI. Indeed there is evidence that children exhibit more severe intellectual impairment than adolescents, suggesting greater vulnerability when diffuse brain insult occurs at an early age. The pattern of findings suggests that visuospatial impairment persists even when verbal ability recovers.

5.2.5. Other Specific Neuropsychological Sequelae

Few studies have assessed neuropsychological sequelae other than global intellectual ability. Klonoff and his colleagues (Klonoff et al., 1977; Klonoff and Low, 1974) administered a comprehensive series of tests which included techniques developed by Reitan, Benton, and Klove to 231 pediatric CHI cases in whom mild injuries predominated. Initial and one-year follow-up findings disclosed that slow visuomotor performance on the Trailmaking Test, reduced finger- and foot-tapping speed, impaired formboard assembly, and defective maze performance were among the most marked deficits when head-injured patients were compared with neurologically intact children of similar demographic background.

Similar to the effects of head injury on adults, visuospatial impairment can occur after head injury in children. Levin and Eisenberg (1979a,b) found that visuospatial deficit, as measured by construction of block designs using three dimensional models (Benton and Fogel, 1962) and copying the Bender designs, was present in nearly one third of pediatric CHI patients tested after resolution of PTA. The likelihood of visuospatial impairment was directly related to the duration of coma and to the severity of acute injury as reflected by ratings on the Glasgow Coma Scale at the time of admission. The authors also found evidence of diminished somatosensory performance on tests of stereognosis, finger localization, and graphesthesia in about one fourth of the total series. Information about the persistence of these sequelae awaits further study. It may be concluded, however, that children exhibit a range of neuropsychological deficits during the first six months after CHI which resembles that found in adults.

5.3. Residual Behavioral Disturbance after Head Injury in Children

It has been shown that head injury producing a long period of coma and acute neurologic deficit frequently results in major behavioral changes in children. In a study of 23 predominantly preschool-aged CHI patients who had been comatose for 24 hr or more, Hjern and Nylander (1964) found that severe psychiatric sequelae were present at least six months after injury in five patients and were accompanied by neurologic deficit in all these cases. The characteristics of the behavioral disorder were not described in detail. Children in coma at least one week were examined one to seven years later and found by Brink and colleagues (1970) to exhibit persistent behavioral changes which varied according to age at injury; children who were under ten years of age at the time of injury exhibited hyperactivity, short attention span, impulsiveness, and aggressive behavior, whereas patients who were at least ten years old at the time of injury manifested poor judgement and affective disturbance.

A follow-up study by Flach and Malmros (1972), which was completed eight to ten years after CHI of varying severity, found that 27% of their patients were "socially maladjusted." Severe complications of acute injury, such as cardiac arrest and respiratory failure, were related to chronic social problems in this study. The authors depicted a characteristic "slowness" in mental function and motor behavior, but were unable to confirm the observations of hyperkinesis reported by Brink *et al.* Ten children and adolescents who had periods of PTA for at least a week were subsequently found by Richardson (1963) to require a "sheltered and tolerant environment" and to frequently manifest perseverative behavior associated with increased anxiety, irritability, and excessive fatigability.

In a recent prospective study of residual behavioral disturbance two years after head injury of varying severity in children, Brown and colleagues (1981) found that one half of the severely injured (PTA > 7 days) patients developed a psychiatric disorder which had not been present before. Based on a psychiatric examination and interviews with parents and teachers, the authors characterized the behavioral disorder after severe head injury as a "disinhibited state," i.e., restlessness with inappropriate or profane remarks, impulsive, sexually explicit behavior, and carelessness in personal hygiene. Consistent with the earlier study by Flach and Malmros (1972), Brown and colleagues found no rise in hyperkinesis after head injury. Children with less severe head injury (PTA < 1 week) more frequently exhibited behavioral disturbance before the accident as compared with a control group, but there was no appreciable behavioral change postinjury. Age at injury was not related to residual psychiatric disturbance. In summary, there is a consensus that development of psychiatric disorder in head-injured children without preexisting behavioral disturbance is confined to patients who sustain severe brain injury.

The findings of these studies resemble Blau's 1936 description of the "post-traumatic psychopathic personality" in which the cardinal characteristics were overactivity, restlessness, destructiveness, aggression, temper tantrums, and delinquency. From serial observations, Blau inferred that behavioral disorder attributable to brain injury became progressively apparent after convalescence from the acute injury when the patient engaged in socially unacceptable behavior that brought him into conflict with the community.

Behavioral sequelae one to two years after CHI of predominantly mild to moderate severity were systematically evaluated by Klonoff and Low (1974) on the basis of behavioral observations and parental interviews. Males were also found to be more vulnerable to developing behavioral disturbance after CHI; increased irritability was commonly observed in young boys at both the one- and two-year postinjury examinations, whereas "personality changes" were frequently reported in both sexes irrespective of age. Overall, slightly more than one half of the total series ($N = 231$) evidenced at least one residual behavioral problem two years after injury. Notwithstanding the alleged rare occurrence of the "posttraumatic syndrome" in children (Jennett, 1972), 29% of the children studied by Klonoff and Low complained of headache during the first year and 18% continued to complain of headache during the second year after injury. Dizziness (vertigo, giddiness) was common in boys during the first year after injury, but greatly diminished in frequency thereafter.

Several investigators have analyzed the potentially moderating effects of age, severity and locus of injury, premorbid factors, and social conditions on the emergence of behavioral disturbance after head injury. In a comprehensive analysis of age effects, Klonoff and Low (1974) found no difference in the relative frequency of sequelae and only minor variation in the rank order of problems when children who were under nine years of age at the time of injury were compared with older CHI patients. Similarly, Shaffer et al. (1975) reported that age had no bearing on behavioral problems resulting from CHI with unilateral depressed skull fracture. The authors reported that severity of CHI, as measured by duration of coma, was related to the likelihood of later psychiatric disturbance, though the lateralization of depressed skull fracture was inconsequential.

5.4. Summary of Research on Recovery after Head Injury

The previously held view that children are relatively impervious to cognitive impairment after CHI is clearly not supported by the available data. Several studies have documented persistent, marked cognitive impairment concomitant with a decline in school achievement after severe injury in children. Differentiation of the effects of primary brain damage of the immediate-impact type from secondary-brain injury associated with late complications (e.g., hypoxia, metabolic disturbance) is necessary to elucidate the relationship between mechanisms of injury and neuropsychological sequelae.

The evidence suggests that memory and motor skill are the most seriously impaired functions after CHI in children. In contrast, language development is relatively resistant to the effects of head injury as reflected by predominantly mild aphasic defects and their remarkable reversibility. Behavioral disturbance has been demonstrated to occur after severe CHI in children, though a consensus has not been reached concerning the most salient features of post-CHI personality changes. The relative contributions of brain damage, preinjury characteristics, and the postinjury social milieu to behavioral disturbance remain obscure.

6. DIRECTIONS FOR FUTURE RESEARCH

The results of recent animal experiments and findings in studies of recovery from brain injury in infants and children are compatible in that they question any sweeping generalization about greater functional plasticity as compared with similar injuries to the mature brain. More research is necessary to develop a scale of severity of acute brain injury in children that would have similar prognostic implications as the Glasgow Coma Scale score in adults.

While the findings in animal studies have demonstrated delayed onset of a performance deficit after an apparently satisfactory recovery, this phenomenon warrants closer study in young children who sustain brain injury. Comparative study of recovery from early injury would also be facilitated by administration of parallel tasks. Development of a delayed spatial alternation test for children with focal frontal lobe injury would permit a comparison with the experimental findings reported by Goldman and her colleagues (1974, 1977).

The results of hemispherectomy in infants and young children indicate that the right hemisphere can provide an adequate substrate for language, even when speech is established before surgical removal of the left hemisphere. It is also clear that children recover more rapidly, if not more fully, from acquired aphasia than adults. However, even children exhibit residual linguistic defects to a greater degree than previously realized. To what extent the right hemisphere increases its participation in language after left-hemisphere injury in young children remains unexplored. Noninvasive physiological techniques are now available which can assess hemispheric activation on linguistic and nonverbal tasks and thus evaluate the role of the right hemisphere in language after early left-hemisphere injury.

The outcome of diffuse brain injury is decidedly worse in young children as compared with older children and young adults. However, further research is necessary to elucidate the pathophysiologic mechanisms which contribute to the relatively poor recovery after early diffuse brain injury. The new techniques of nuclear magnetic resonance and positron emission tomography could provide information on the relationship between alteration in cerebral metabolism and cognitive sequelae after diffuse injury.

Memory deficit is the most common sequel of head injury in children. Only a few studies, however, have considered different types of memory (e.g., recall versus recognition) or have attempted to assess the effects of brain injury on various stages of information processing (e.g., encoding, storage, retrieval). Most of the existing memory tests employed in the pediatric age group are modificaions of procedures that were designed to study adults with brain damage. Consequently, there is a need to develop tasks that are more appropriate and more interesting for the young child. Moreover, there is no standardized technique for assessing acute disorientation and confusion in young children.

In conclusion, the availability of noninvasive physiological techniques and improved neuropsychological tests to evaluate subtle language disturbance, memory deficit, and other sequelae hold considerable promise for improving our understanding of recovery from early brain injury. These methods of investigation might also suggest techniques for educational and behavioral remediation of impairments that frequently result after brain injury in children.

ACKNOWLEDGMENTS. This work was supported in part by NS 077377-11, Center for the Study of Nervous System Injury Grant, Contract NS 9-2314, Comprehensive Central Nervous System Trauma Center, and Moody Foundation 80-233. We are indebted to Lori Bertolino, Sarah De Los Santos, Pat Smith, and Karen St. Claire for assistance in preparation of this manuscript and data analysis.

REFERENCES

Alajouanine, Th., and Lhermitte, F., 1965, Acquired aphasia in children, *Brain* **88**:653–662.

Annegers, J. F., Grabow, J. D., Kurland, L. T., and Laws, E. R., 1980, The incidence, causes, and secular trends of head trauma in Olmsted County, Minnesota, *Neurology* **30**:912–919.

Annett, M., 1973, Laterality of childhood hemiplegia and the growth of speech and intelligence, *Cortex* **9**:4–33.

Basser, L. S., 1962, Hemiplegia of early onset and the faculty of speech with special reference to the efects of hemispherectomy, *Brain* **85**:427–460.

Becker, D. P., Miller, J. D., Ward, J. D., Greenberg, R. P., Young, H. F., and Sakalas, R., 1977, The outcome from severe head injury with early diagnosis and intensive management, *J. Neurosurg.* **47**:491–502.

Bell, E., and Karnosh, L. J., 1949, Cerebral hemispherectomy: Report of a case ten years after operation, *J. Neurosurg.* **6**:285–293.

Benton, A. L., and Fogel, M. L., 1962, Three dimensional constructional praxis: A clinical test, *Arch. Neurol.* **7**:347–354.

Bishop, D. V., 1981, Plasticity and specificity of language localization in the developing brain, *Dev. Med. Child Neurol.* **23**:251–255.

Blau, A., 1936, Mental changes following head trauma in children, *Arch. Neurol. Psychiatr.* **35**:723–769.

Brink, J. D., Garrett, A. L., Hale, W. R., Woo-Sam, J., and Nickel, V. L., 1970, Recovery of motor and intellectual function in children sustaining severe head injuries, *Dev. Med. Child Neurol.* **12**:565–571.

Brink, J. D., Imbus, C., and Woo-Sam, J., 1980, Physical recovery after severe closed head trauma in children and adolescents, *J. Pediatr.* **97**:721–727.

Brown, G., Chadwick, O., Shaffer, D., Rutter, M., and Traub, M., 1981, A prospective study of children with head injuries: III. Psychiatric sequelae, *Psychol. Med.* **11**:63–78.

Bruce, D. A., Schut, L., Bruno, L. A., Wood, H. H., and Sutton, L. N., 1978, Outcome following severe head injury in children, *J. Neurosurg.* **48**:679–688.

Bruell, J. H., and Albee, G. W., 1962, Higher intellectual functions in a patient with hemispherectomy for tumors, *J. Consult. Clin. Psychol.* **26**:90–98.

Burkland, C. W., and Smith, A., 1977, Language and the cerebral hemispheres, *Neurology* **27**:627–633.

Buschke, H., and Fuld, P. A., 1974, Evaluating storage, retention, and retrieval in disordered memory and learning, *Neurology* **24**:1019–1025.

Carlson, J., Netley, C., Hendrik, E. B., and Pritchard, R. S., 1968, A re-examination of intellectual disabilities in hemispherectomized patients, *Trans. Am. Neurol. Assoc.* **93**:198–201.

Chadwick, O., Rutter, M., Shaffer, D., and Shrout, P. E., 1981, A prospective study of children with head injuries: IV. Specific cognitive deficits, *J. Clin. Neuropsychol.* **3**:101–120.

Cotard, J., 1968, Etude sur l'atrophie partielle du cerveau, Thèse de Paris.

Craft, A. W., 1972, Head injury in children in: *Handbook of Clinical Neurology* Volume 23 (P. J. Vinken and G. W. Bruyn, eds.), Elsevier North-Holland, New York, pp. 445–458.

Damasio, A. R., Lima, A., and Damasio, H., 1975, Nervous function after right hemispherectomy, *Neurology* **25**:89–93.

Davidson, P. W., Willoughby, R. H., O'Tuama, L. A., Swisher, C. N., and Benjamins, D., 1978, Neurological and intellectual sequelae of Reye's syndrome, *Am. J. Ment. Defic.* **82**:535–541.

Day, P. S., and Ulatowska, H. K., 1979, Perceptual, cognitive, and linguistic development after early hemispherectomy: Two case studies, *Brain Lang.* **7**:17–33.

Dennis, M., and Whitaker, H. A., 1976, Hemispheric equipotentiality and language acquisition, in: *Language Development and Neurologic Theory* (S. Segalowitz and F. Gruber, eds.), Academic Press, New York, pp. 93–104.

Dikmen, S., Matthews, C. G., and Harley, J. P., 1975, The effect of early versus late onset of major motor epilepsy upon cognitive-intellectual performance, *Epilepsia* **16**:73–81.

Duffy, T. E., Cavazzuti, M., Cruz, N. F., and Sokoloff, L., 1982, Local cerebral glucose metabolism in newborn dogs: Effects of hypoxia and halothane anesthesia, *Ann. Neurol.* **11**:233–246.

Eidelberg, E., and Stein, D. G., 1974, Functional recovery after lesions of the nervous system, *Neurosci. Res. Prog. Bull.* **12**:195–233.

Eiser, C., 1978, Intellectual abilities among survivors of childhood leukaemia as a function of CNS irradiation, *Arch. Dis. Child.* **53**:391–395.

Field, J. H. (ed)., 1976, *Epidemiology of Head Injury in England and Wales: With Particular Application to Rehabilitation,* printed for H. M. Stationery Office by Willsons, Leicester.

Finger, S. (ed.), 1978, *Recovery from Brain Damage: Research and Theory,* Plenum Press, New York.

Flach, J., and Malmros, R., 1972, A long-term follow-up study of children with severe head injury, *Scand. J. Rehabil. Med.* **4**:9–15.

Fuld, P. A., and Fisher, P., 1977, Recovery of intellectual ability after closed head-injury, *Dev. Med. Child. Neurol.* **19**:495–502.

Gaidolfi, E., and Vignolo, L. A., 1980, Closed head injuries of school-age children: Neuropsychological sequelae in early adulthood, *Ital. J. Neurol. Sci.* **1**:65–73.

Gardner, W. J., Karnosh, L. J., McLure, C. C., and Gardner, A. K., 1955, Residual function following hemispherectomy for tumour and for infantile hemiplegia, *Brain* **78**:487–502.

Gillies, J. D., and Seshia, S. S., 1980, Vegetative state following coma in childhood: Evolution and outcome, *Dev. Med. Child Neurol.* **22**:642–648.

Goldman, P. S., 1974, An alternative to developmental plasticity: Heterology of CNS structures in infants and adults, in: *Plasticity and Recovery of Function in the Central Nervous System* (D. G. Stein, J. J. Rosen, and N. Butters, eds.), Academic Press, New York, pp. 149–174.

Goldman, P. S., and Alexander, G. E., 1977, Maturation of prefrontal cortex in the monkey revealed by focal reversible cryogenic depression, *Nature* **267**:613–615.

Gott, P. S., 1973a, Language after dominant hemispherectomy, *J. Neurol. Neurosurg. Psychiatr.* **36**:1082–1088.

Gott, P. S., 1973b, Cognitive abilities following right and left hemispherectomy, *Cortex* **9**:266–274.

Graham, F. K., Ernhart, C. B., Thurston, D., and Craft, M., 1962, Development three years after perinatal anoxia and other potentially damaging newborn experiences, *Psychol. Monogr.: Gene. Appl.* **76**(3):1–53.

Griffith, H., and Davidson, M., 1966, Long-term changes in intellect and behavior after hemispherectomy, *J. Neurol. Neurosurg. Psychiatr.* **29**:571–576.

Gurdjian, E. S., and Webster, J. E., 1958, *Head Injuries: Mechanisms, Diagnosis, and Management*, Little, Brown, Boston.

Guttman, E., 1942, Aphasia in children, *Brain* 65:205–219.

Hannay, H. J., Levin, H. S., and Grossman, R. G., 1979, Impaired recognition memory after head injury, *Cortex* 15:269–283.

Hécaen, H., 1976, Acquired aphasia in children and the ontogenesis of hemispheric functional specialization, *Brain Lang.* 3:114–134.

Heiskanen, O., and Kaste, M., 1974, Late prognosis of severe brain injury in children, *Dev. Med. Child Neurol.* 16:11–14.

Hendrick, E. B., Harwood-Hash, D. C. F., and Hudson, A. R., 1964 Head injuries in children: A survey of 4465 consecutive cases at the Hospital for Sick Children, Toronto, Canada, *Clin. Neurosurg.* 11:46–59.

Hjern, B., and Nylander, I., 1964, Late prognosis of severe head injuries in childhood, *Acta Paediatr. Scand. (Suppl.)* 152:113–116.

Ignelzi, R. J., and Bucy, P. C., 1968, Cerebral hemidecortication in the treatment of infantile cerebral hemiatrophy, *J. Nerv. Ment. Dis.* 147:14–30.

Ivnik, R. J., Colligan, R. C., Obetz, S. W., Smithson, W. A., 1981, Neuropsychological performance among children in remission from acute lymphocytic leukemia, *Dev. Behav. Pediatr.* 2:29–34.

Jennett, B., 1972, Head injuries in children, *Dev. Med. Child Neurol.* 14:137–147.

Jennett, B., and Plum, F., 1972, Persistent vegetative state after brain damage, *Lancet* 1:734–737.

Jennett, B., and Teasdale, G., 1981, *Management of Head Injuries*, F. A. Davis, Philadelphia.

Jennett, B., Teasdale, G., Galbraith, S., Pickard, J., Grant, H., Braakman, R., Avezaat, C., Maas, A., Minderhoud, J., Vecht, C. J., Heiden, J., Small, R., Caton, W., and Kurtz, T., 1977, Severe head injuries in three countries, *J. Neurol. Neurosurg. Psychiatr.* 40:291–298.

Kennard, M. A., 1938, Reorganization of motor functions in the cerebral cortex of monkeys deprived of motor and premotor areas in infancy, *J. Neurophysiol.* 1:477–497.

Klonoff, H., 1971, Head injuries in children: Predisposing factors, accident conditions, accident proneness, and sequelae, *Am. J. Publ. Health* 61:2405–2417.

Klonoff, H., and Low, M., 1974, Disordered brain function in young children and early adolescents: Neuropsychological and electroencephalographic correlates, in: *Clinical Neuropsychology: Current Status and Applications* (R. M. Reitan and L. A. Davison, eds.), John Wiley and Sons, New York, pp. 121–178.

Klonoff, H., Low, M. D., and Clark, C., 1977, Head injuries in children: A prospective five year followup, *J. Neurol. Neurosurg. Psychiatr.* 40:1211–1219.

Knobloch, H., and Pasamanick, B., 1959, Syndrome of minimal cerebral damage in infancy, *J. Am. Med. Assoc.* 170:1384–1387.

Kohn, B., and Dennis, M., 1974, Somatosensory functions after cerebral hemidecortication for infantile hemiplegia, *Neuropsychologia* 12:119–130.

Kraus, J. F., 1980, Injury to the head and spinal cord: The epidemiological relevance of the medical literature published from 1960 to 1978, *J. Neurosurg. (Suppl.)* 53:3–10.

Lange-Cosack, H., Wider, B., Schlesner, H.-J., Grumme, Th., and Kubicki, St., 1979, Prognosis of brain injuries in young children (one until five years of age), *Neuropaediatrie* 10:105–127.

Lenneberg, E., 1967, *Biological Foundations of Language*, John Wiley and Sons, New York.

Levin, H. S., and Eisenberg, H. M., 1979a, Neuropsychological outcome of closed head injury in children and adolescents, *Child's Brain* 5:281–292.

Levin, H. S., and Eisenberg, H. M., 1979b, Neuropsychological impairment after closed head injury in children and adolescents, *J. Pediatr. Psychol.* 4:389–402.

Levin, H. S., and Grossman, R. G., 1976, Effects of closed head injury on storage and retrieval in memory and learning of adolescents, *J. Pediatr. Psychol.* 1:38–42.

Levin, H. S., Grossman, R. G., and Kelly, P. J., 1976, Aphasic disorder in patients with closed head injury, *J. Neurol. Neurosurg. Psychiatr.* 39:1062–1070.

Levin, H. S., Eisenberg, H. M., Wigg, N. R., and Kobayashi, K., 1982, Memory and intellectual ability after head injury in children and adolescents, *Neurosurgery* 11:668–673.

Lindsay, J., Ounsted, C., and Richards, P., 1979, Long-term outcome in children with temporal lobe seizures. I: Social outcome and childhood factors, *Dev. Med. Child Neurol.* 21:285–298.

McFie, J., 1961, The effects of hemispherectomy on intellectual functioning in cases of infantile hemiplegia, *J. Neurol. Neurosurg. Psychiatr.* 24:240–249.

Meadows, A. T., Massary, D. J., Fergusson, J., Gordon, J., Littman, P., Moss, K., 1981, Declines in IQ scores and cognitive dysfunctions in children with acute lymphosytic leukaemia treated with cranial irradiation, *Lancet* 2:1015–1018.

Mensh, I. N., Schwartz, H. G., Matarazzo, R. R., and Matarazzo, J. D., 1952, Psychological functioning following cerebral hemispherectomy in man, *Arch. Neurol. Psychiatr.* 67:787–796.

Moyes, C. D., 1980, Epidemiology of serious head injuries in childhood, *Child Care, Health Dev.* 6:1–9.

Norman, M. G., 1978, Perinatal brain damage, *Perspect. Pediatr. Pathol.* 4:41–92.

O'Leary, D. S., Seidenberg, M., Berent, S., and Boll, T. J., 1981, Effects of age of onset of tonic-clonic seizures on neuropsychological performance in children, *Epilepsia* 22:197–204.

Richardson, F., 1963, Some effects of severe head injury. A follow-up study of children and adolescents after protracted coma, *Dev. Med. Child Neurol.* 5:471–482.

Rowbotham, G. F., Maciver, I. N., Dickson, J., and Bousfield, M. E., 1954, Analysis of 1,400 cases of acute injury to the head, *Br. Med. J.* 1:726–730.

Rudel, R. G., and Teuber, H.-L., 1971, Spatial orientation in normal children and in children with early brain injury, *Neuropsychologia* 9:401–407.

Rudel, R. G., Teuber, H.-L., and Twitchell, T. E., 1974, Levels of impairment of sensory-motor functions in children with early brain damage, *Neuropsychologia* 12:95–108.

Satz, P., and Bullard-Bates, C., 1981, Acquired aphasia in children, in: *Acquired Aphasia* (M. T. Sarno, ed.), Academic Press, New York, pp. 401–423.

Seshia, S. S., Seshia, M. M. K., and Sachdeva, R. K., 1977, Coma in childhood, *Dev. Med. Child Neurol.* 19:614–628.

Shaffer, D., Chadwick, O., and Rutter, M., 1975, Pychiatric outcome of localized head injury in children, in: *Outcome of Severe Damage to the Central Nervous System*, Ciba Foundation Symposium 34 (new series), Elsevier, Excerpta Medica, North-Holland, Amsterdam, pp. 191–213.

Smith, A., 1966, Speech and other functions after left (dominant) hemispherectomy, *J. Neurol. Neurosurg. Psychiat.* 29:467–471.

Smith, A., 1974, Dominant and nondominant hemispherectomy, in: *Hemisphere Disconnection and Cerebral Function* (M. Kinsbourne and W. L. Smith, eds.), Volume II, Charles C. Thomas, Springfield, Illinois, pp. 5–33.

Smith, A., and Sugar, O., 1975, Development of above normal language and intelligence 21 years after left hemispherectomy, *Neurology* 25:813–818.

Spreen, O., and Benton, A. L., 1969, *Neurosensory Center Comprehensive Examination for Aphasia: Manual of Directions*, Neuropsychology Laboratory, University of Victoria, Victoria, B. C.

St. James-Roberts, I., 1979, Neurological plasticity, recovery from brain insult, and child development, *Adv. Child Dev. Behav.* 14:253–319.

Teasdale, G., Skene, A., Parker, L., and Jennett, B., 1979, Age and outcome of severe head injury, *Acta Neurochir. (Suppl.)* 28:140–143.

Todorow, S., 1975, Recovery of children after severe head injury: Psychoreactive superimpositions, *Scand. J. Rehabil. Med.* **7**:93–96.

Towbin, A., 1969, Mental retardation due to germinal matrix infarction, *Science* **164**:156–161.

van Dongen, H. R., and Loonen, M. C. B., 1977, Factors related to prognosis of acquired aphasia in children, *Cortex* **13**:131–136.

Wada, J. A., 1974, Morphologic asymmetry of human cerebral hemispheres: Temporal and frontal speech zones in 100 adult and 10 infant brains, *Neurology* **24**:349.

Wilson, P. J., 1970, Cerebral hemispherectomy for infantile hemiplegia. A report of 50 cases, *Brain* **93**:147–180.

Winick, M., 1976, *Malnutrition and Brain Development,* Oxford University Press, New York.

Witelson, S., and Pallie, W., 1973, Left hemisphere specialization for language in the newborn: Neuroanatomical evidence of asymmetry, *Brain* **96**:641–646.

Woods, B. T., 1980, The restricted effects of right-hemisphere lesions after age one: Wechsler test data, *Neuropsychologia* **18**:65–70.

Woods, B. T., and Carey, S., 1979, Language deficits after apparent clinical recovery from childhood aphasia, *Ann. Neurol.* **6**:405–409.

Woods, B. T., and Teuber, H.-L., 1978, Changing patterns of childhood aphasia, *Ann. Neurol.* **3**:273–280.

Wright, L., 1978, A method for predicting sequelae to meningitis, *Am. Psychol.* **33**:1037–1039.

Yaar, I., Ron, E., Modan, M., Peretz, H., and Modan B., 1979, Long-term cerebral effects of small doses of x-irradiation in childhood as manifested in adult visual evoked responses, *Ann. Neurol.* **8**:261–268.

10

Recovery of Function in Senile Dementia of the Alzheimer Type

STEVEN T. DeKOSKY

1. INTRODUCTION

From the standpoint of the clinical investigator, there appears to be no more difficult issue to address than recovery of function in diseases or processes which progressively deplete brain and behavioral function. Age or disease-associated changes in behavior might be compensated or alleviated in a variety or combination of ways, however. These include behavioral-strategy alterations, pharmacologic manipulations, administration of growth or trophic factors, and transplantation of neural cells into a donor brain, in either an orthotopic or heterotopic location. Each of these shall be discussed briefly, utilizing as a model the clinical entity of senile dementia of the Alzheimer type (AD). We shall speculate on the use of these various strategies to palliate or reverse the symptoms or biochemical changes associated with advanced age or Alzheimer's disease.

2. ALZHEIMER'S DISEASE: THE CLINICAL AND BIOCHEMICAL PROBLEM

Alzheimer's disease is an age-related disorder of cognition and memory, which occurs with increasing frequency above the age of 65 and takes a tremendous social, psychological, financial, and emotional toll upon both patient and

STEVEN T. DeKOSKY • Department of Neurology, Lexington Veterans Administration and University of Kentucky Medical Centers, and Sanders–Brown Research Center on Aging, Lexington, Kentucky 40536.

family. Characterized early in its course by loss of recent memory and altered judgment and insight, it progresses to loss of virtually all memory, cognitive function, and social interaction (Schneck et al., 1982). Described in 1907 in a patient of 52 years of age by Alois Alzheimer (the disease is now believed to be the same whether manifesting itself in the presenium or after age 65), the process is further characterized by cortical atrophy, neurofibrillary tangles in neuronal cell bodies of the cortex and hippocampus, senile plaques composed of degenerating neurites, and granulovacuolar degeneration in the pyramidal cells of the hippocampus. Evidence for a biochemical defect surfaced in the 1970s, when several laboratories (Bowen et al., 1976; Davies and Maloney, 1976; Perry et al., 1977; White et al., 1977) found deficiency of the cholinergic synthetic enzyme choline acetyltransferase (CAT) in the brains of patients with Alzheimer's disease. Significantly, the loss of CAT activitiy was most marked in cortex and hippocampus. Subsequently, the level of this chemical deficiency was shown to correlate with the severity of the dementia (Perry et al., 1978). Therefore, in the late 1970s and into the 1980s, research focused upon the anatomy and pharmacology of the cholinergic neurotransmission system to understand its potential role in Alzheimer's disease.

Several different projection systems or local neural nets use acetylcholine as their neurotransmitter. Projections from large cells in the medial septal nucleus and the diagonal band of Broca in the forebrain supply the hippocampus, which is of fundamental importance in memory processing in animals and humans. Lesion or loss of these cells or section of their projection to the hippocampus, the fimbria–fornix, result in loss of CAT activity in that structure (Lewis et al., 1967). Large cholinergic cells in the basal forebrain, in the ventral pallidum of the rat, and in higher species in the nucleus basalis of Meynert (NBM) appear to supply the cortex of rodents and primates with a large amount of their cholinergic activity (Mesulam, 1983; Price et al., 1982; Gorry, 1963). Still other cholinergic systems are "intrinsic" and do not appear to project out of their loci, exemplified by the highly concentrated cholinergic activity of the caudate and other basal ganglia. It remains to be determined if a significant number of intrinsic cholinergic neurons are to be found in the cortex (Fibiger, 1982).

Destroying by lesion the source of the "extrinsic" cholinergic input to the rat cortex, the ventral pallidum, causes a decline in CAT activity in the cortex (Johnston et al., 1981; Lehmann et al., 1980; Wenk et al., 1980). Since cortical CAT is low in Alzheimer's disease, the cells of the human ventral forebrain believed to be the source of the CAT to the cortex were examined and found to be greatly decreased in number in Alzheimer's disease (Averback, 1981; Whitehouse et al., 1982). The cells of the nucleus basalis which remain in Alzheimer's disease patients were found to have decreased RNA in the cytoplasm and lower nucleolar volume than those of controls, suggesting decreased perikaryal, as well as presynaptic, dysfunction (Mann and Yates, 1982). Examination of the source of the hippocampal cholinergic innervation, the medial septum and diagonal band, has not yet been reported, but would be expected to show similar changes. At least some of the prominent behavioral and biochemical alterations in the brains of Alzheimer patients may be characterized as a dysfunction of extrinsic cholinergic projection systems (Price *et al.*, 1982). The disease is more complex than that; certain subpopulations of patients may have noradrenergic deficits as well (Bondareff *et al.*, 1982). In addition, neuronal membrane markers and myelin markers are decreased in Alzheimer brains, suggesting a wider loss of connectivity than could be accounted for by loss of cholinergic circuitry alone (DeKosky and Bass, 1982; Bowen *et al.*, 1977; Cherayil, 1969; Buell and Coleman, 1979). However, the correlation of severity of the dementia with cholinergic loss (Perry *et al.*, 1978) was sufficient to initiate attempts to enhance cholinergic neurotransmission in humans, especially since such manipulations improve behavioral performance in aged rodents and monkeys (Bartus *et al.*, 1982).

3. BEHAVIORAL STRATEGIES FOR COGNITIVE DEFICITS OF ALZHEIMER'S DISEASE

Various strategies are adopted by family members, nurses, nursing homes, and other persons dealing with mildly or moderately demented patients. The strategies are different depending upon the degree of cognitive deficit from which an individual patient suffers.

In cases with early, mild memory deficits, cultivation of habits of making lists or immediately writing things down are quite helpful and compensate for faulty recall. Easily available indices of orientation such as large wall clocks and calendars provide time reference and reduce the dementia patients' sense of being disoriented or lost in time.

In patients with more severe cognitive deficits, daily schedules are arranged for consistency and predictability. An expected schedule of environmental events appears comforting, and deviation from such a schedule not infrequently produces worsened disorientation, global confusion, and secondary affective changes.

The disruption of behavioral function resulting from the intrusion of new, strange, or unexpected environmental events may reflect the inability of the impaired neural systems to perform at any heightened level of activity beyond the baseline. Thus overlearned or familiar activities of daily living are performed well, but novel situations requiring activation and integration of attentional, cognitive, and memory processes are not. The extent to which specific neurotransmitter system pathology is linked to these problems remains to be determined. As effective pharmacotherapy directed at the cholinergic system is implemented, the role of the cholinergic deficit in this problem will be better understood.

At the present time, the behavioral strategies employed in all but very early cases of Alzheimer's disease are attempts to preserve an environment where the patient's previously learned skills, including activities of daily living, may be preserved and utilized. It is a mark of the decreased functional reserve of the Alzheimer brain that sudden decreases in cognitive function brought about by an illness, changing living quarters, or death of a spouse take a long time to recover to the previous prestressed baseline. Specific behavioral strategies to be developed will likely involve state-dependent learning paradigms, wherein an effective pharmaceutical, improving some specific neural function, can be used in combination with behavioral techniques to provide an improved quality of life for both the patient and his caretakers.

4. PHARMACOLOGIC INTERVENTION IN ALZHEIMER'S DISEASE

The focus of attention on cholinergic transmission was heightened by the earlier demonstration of interference with normal human memory processes by cholinergic blockers such as scopolamine (Drachman and Leavitt, 1974). Since memory loss is the *sine qua non* of Alzheimer's disease, the concept of cholinergic system dysfunction as largely responsible for that behavioral defect appeared promising. A lack of quantitative alteration of the muscarinic cholinergic receptor (Perry *et al.*, 1977; White *et al.*, 1977; Davies and Verth, 1978; Bowen *et al.*, 1979) indicated that the cholinergic problem might be "presynaptic" and suggested that functional restoration was possible with pharmacologic agents directed at stimulating the spared cholinergic receptor directly or indirectly.

The search for effective pharmacologic treatment based on the cholinergic neurotransmission deficit has focused upon multiple areas of potential intervention along the cholinergic synaptic pathway. Precursor loading of choline or lecithin to increase acetylcholine at the synapse has been tried with little success, although occasional improvements have been noted in individuals. Physostigmine, an acetylcholinesterase inhibitor, has been used with only mild success in both humans and animals to improve memory function, presumably by delaying the degradation of acetylcholine and extending its modulatory effects (Bartus *et al.*, 1982). Direct stimulation of the postsynaptic receptor by a cholinergic agonist (arecholine) enhances memory function in young normal humans (Sitaram *et al.*, 1978). In Alzheimer patients it is reported to show similar mild effects (Christie *et al.*, 1981). Combinations of a cholinergic precursor such as choline and a stimulant of oxidative metabolism (piracetam) were noted to be beneficial for some individuals in a study of Alzheimer patients. Biochemical differences were apparent in red cell choline of "responders" versus "nonresponders" (Friedman *et al.*, 1981).

Recently naloxone, an opiate receptor blocker, has been shown to induce transient improvement in performance of memory tasks by patients with Alzheimer's disease when the drug was administered intravenously (Reisberg *et al.*, 1983). The approach was based upon the known involvement of opiate neurotransmitter systems in learning and memory. The patient's improvement might involve blocking the suppressive effects of opiates on acetylcholine metabolism (Moroni *et al.*, 1978), thus disinhibiting the remaining cholinergic neurons in the basal forebrain. The inhibitory modulating effects of opiates on brain acetylacholine turnover can be blocked by naloxone (Moroni *et al.*, 1978). Such a mechanism would also be consistent with the cholinergic defect hypothesis of Alzheimer's disease.

It is notable that, from data such as are available, combinations of neurotransmitter-related substances and metabolic enhancers such as piracetam are thus far of most help. However, no therapy has succeeded in significantly improving function or quality of life in these patients. Complete and thoughtful examinations of the "cholinergic hypothesis" of memory and cholinergic dysfunction in Alzheimer's disease and aging discuss these issues in more detail (Crook and Gershon, 1981; Bartus *et al.*, 1982). Efforts to replace the missing neurotransmitter hinge on the presence of the postsynaptic receptor site. As noted earlier, cholinergic receptor binding sites appear to be of normal density in Alzheimer's disease. However, while CAT has a specific laminar distribution in cortex, human frontal cortical cholinergic receptor binding is equal throughout the layers (DeKosky *et al.*, 1983b). A strong correlation between presynaptic and postsynaptic cholinergic markers is not demonstrable in human or in rat cortex (Fibiger, 1982; DeKosky *et al.*, 1983a).

5. CHOLINERGIC SYSTEMS, TROPHIC SUBSTANCES, AND POTENTIAL THERAPIES FOR ALZHEIMER'S DISEASE

An hypothesis of the etiology of Alzheimer's disease involving a missing trophic factor has been advanced by Appel (1981). Making specific reference to the hippocampal and cortical cholinergic deficits from what is likely impaired extrinsic innervation, Appel suggested that lack of a neurotrophic factor secreted by the innervated cortical and hippocampal cells might be responsible for the dying back of projections from the basal forebrain cells (Basal nucleus of Meynert, nucleus of the diagonal band, and medial septum). He extended the model to two other neurological diseases of uncertain etiology—amyotrophic lateral sclerosis and Parkinson's disease, where the necessary but missing trophic substances are produced by muscle and basal ganglia cells, respectively. Such lack of the trophic factor might lead to loss of the axonal distribution (to cortex and hippocampus in the case of Alzheimer's disease) and lead to cholinergic and subsequent behavioral deficits.

Evidence for factors exerting trophic effects on cholinergic systems has begun to emerge in the past several years (Helfand *et al.*, 1976; Varon *et al.*, 1981; Barde *et al.*, 1983). Isolation of the factors and efforts to assess their in vitro and in vivo effects on neurons have accelerated since Helfand *et al.* (1976) reported that a soluble factor from medium conditioned with chick embryo heart cells (HCM) promoted survival of ciliary ganglion neurons in culture. The factor also stimulated vigorous neuritic outgrowth of the cultured cholinergic neurons (Collins, 1978). Cholinotrophic activity has also been described from extracts of eye from 15-day chick embryo (Manthorpe *et al.*, 1980). In vitro studies suggested that the neurotrophic effects of the neuritic outgrowth effects were separable—perhaps indicating two different proteins (Varon *et al.*, 1981; Adlu and Varon, 1980; Collins, 1978).

A quantitative assay for neuritic outgrowth has been developed (Varon *et al.*, 1981). Heart cell medium is obtainable in large amounts, and bulk separation of the chick embryo eye extract for isolation and characterization of the factors has been reported (Manthorpe *et al.*, 1980).

The first evidence of in vivo regeneration of central cholinergic pathways using a trophic agent indicated that HCM, when administered through a cannula to the cells of the medial septum, accelerated new growth of cholinergic fibers into an iris transplanted into the hippocampus of an adult rat (Schonfeld *et al.*, 1981). Thus neuritic (axonal) outgrowth was influenced in vivo by one of the trophic factors, cholinergic neurotrophic factor (CNTF). This trophic effect was elicited in a different species (rat) from that from which the factor was derived (chick), a fact important for potential therapeutic applications. How might such a factor or factors be directed to reverse or at least palliate the loss of cholinergic innervation in humans with Alzheimer's disease?

Further demonstrations of the axonal outgrowth effects of the CNTF need to be demonstrated in vivo, such as increasing cortical or hippocampal activity in an appropriate animal model. The target organ in the initial study was an iris; it must be demonstrated that CNTF can induce reinnervation (or hyperinnervation) of its appropriate target cells in vivo.

Route of administration of the CNTF is also a factor, not only because of the potential clinical applications but also for correct interpretation of experimental results. Direct administration of the factor onto the cells of interest worked in the study by Schonfeld *et al.* (1981). Parenteral administration is not practical since the compound is likely of large molecular weight (Manthorpe *et al.*, 1980) and would not be expected to cross the blood–brain barrier in significant amounts.

Experiments performed with nerve growth factor (NGF), the first discovered and best-characterized neurotrophic factor, are instructive. Nerve growth factor exerts a trophic effect on sympathetic neurons and adrenergic sprouting. Direct injection of NGF onto the cells of the local coeruleus was quite effective in stimulating adrenergic sprouting into denervated iris which had been implanted in rat diencephalon (Stenevi *et al.*, 1974). Intraventricular administration of NGF also had significant effects on regeneration (Bjerre *et al.*, 1973), although ventricular NGF administration required concentrations 100 times higher than those doses effective with intraparenchymal injection.

Intraventricular, intracisternal (via the cisterna magna), or even intrathecal (lumbar sac) installation of such a cholinergic factor would be preferable to intraparenchymal administration in man, not only because of the technical difficulties of cannulation, but also because of the location, in man and other primates, of the NBM. In higher mammals the cholinergic cells become scattered in clusters in the basal forebrain (Gorry, 1963; Price *et al.*, 1982) and correct placement of a cannula would be quite difficult. It would be less difficult but still formidable to locate in the medial septum or cells of the diagonal band of Broca. From a practical standpoint, the administration of such factors through a ventricular catheter via an Ommaya reservoir would be the most efficacious way of delivering such a substance.

Toxicity of such administered compounds and the half-life of such factors in the CSF would need to be determined to prescribe frequency of dosing. In studies of NGF where lesions have been made and the effects of NGF on adrenergic sprouting assessed, one administration of NGF was employed and no further injections were helpful (Bjerre *et al.*, 1973). Peptidases or proteinases might inactivate CNTF and alter the effective concentration which would eventually diffuse throughout the spinal fluid to target cells of interest. In vitro studies of the compound with CSF and in vivo studies in subhuman primates ought to answer some of these questions. The slow progressive loss of cholinergic innervation in Alzheimer's disease makes a priori judgment of frequency of administration impossible and such scheduling of human dosages would need to be empirical.

If the overwhelming majority of basal forebrain cholinergic neurons have died before a definitive diagnosis of Alzheimer's disease can be made, factors which enhance the neurite outgrowth or CAT-inducing activities of these cells would likely not be effective. At the time of diagnosis enough basal cholinergic neurons need be present to serve as targets of exogenously administered trophic factors. Initial reports of NBM cell density in Alzheimer's disease indicated greater than 90% loss of these cells, and CAT decreased in NBM in AD (Rossor *et al.*, 1982). More recent studies indicate that lesser numbers of these cells are lost in Alzheimer's disease (Tagliavini and Pilleri, 1983) and that significant basal forebrain cholinergic cell survival may still be associated with drastic losses of cortical CAT (Perry *et al.*, 1982). Such pathologic findings suggest a "dying back" phenomenon and offer the possibility of addressing trophic substance efforts toward this remaining population of cholinergic neurons.

If intraventricular administration of CNTF showed a cholinergic regenerative effect in the target tissues (e.g., increasing CAT and cholinergic activity in hippocampus and cortex of an animal model), the specificity of the effect would assume importance. Enhancement of cholinergic metabolism in other, uninvolved systems might lead to autonomic nervous system dysfunction (hypothalamus and descending tracts), movement disorders such as chorea (intrinsic basal ganglia cholinergic circuitry), dyscoordination (cerebellar systems), or other unwanted side effects. Just as dopaminergic antagonists or agonists have side effects in dopaminergic systems other than those for which they are targeted, such side effects of cholinergic "enhancement" may be problematic. To the extent that these other cholinergic systems are relatively uninvolved in the Alzheimer degenerative process, they might be less sensitive to the effects of CNTF and not present overwhelming problems.

6. PHARMACOLOGIC ENHANCEMENT OF A TROPHIC FACTOR EFFECT

Behavioral or functional improvement will be the primary goal in application of the principles of neurologic regeneration to the clinical problems of Alzheimer's disease. Even a small effect on cholinergic neurotransmission might have a beneficial outcome. In rats which undergo section of the fimbria-fornix and transplantation of embryonic septal nuclei into the lesioned area, regeneration of cholinergic axons into the hippocampal targets occurs, although this growth is not accompanied by improvement on a behavioral task. However, if the transplanted animal is also injected with physostigmine, a cholinesterase inhibitor, task performance is significantly improved (Low *et al.*, 1982). Thus pharmacologic amplification of the cholinergic signal is a potential therapeutic strategy whether in combination with CNTF administration or following transplantation of cholinergic cells.

The speculation of this monograph seems premature, yet the methodology is available and studies bearing on this issue are underway in many laboratories. Issues such as trophic factor toxicity, half-life, dosing regimen, side effects, and effectiveness of route of administration must be determined in vitro and in vivo prior to use in human disease. The potential for such trophic agent use in humans, underway already for NGF, may serve as a guide and catalyst for researchers endeavoring to understand basic regenerative mechanisms and allow an approach to recovery of function in Alzheimer's disease.

REFERENCES

Adler, R., and Varon, S., 1980, Cholinergic neurotrophic factors. V. Segregation of survival- and neurite-promoting activities in heart-conditioned media, *Brain Res.* **188:**437–448.

Alzheimer, A., 1907, Über eine eigenartige Erkrankung der Hirnrinde, *All. Z. Psychiatr.* **64:**146–148.

Appel, S. H., 1981, A unifying hypothesis for the cause of amyotrophic lateral sclerosis, parkinsonism, and Alzheimer disease, *Ann. Neurol.* **10:**499–505.

Averback, P., 1981, Lesions of the nucleus ansae peduncularis in neuropsychiatric disease, *Arch. Neurol.* **38:**230–235.

Barde, Y.-A., Edgar, D., and Thoenen, H., 1983, New neurotrophic factors. *Annu. Rev. Physiol.* **45:**601–612.

Bartus, R., Dean, R. L., Beer, B., and Lippa, A., 1982, The cholinergic hypothesis of geriatric memory dysfunction. *Science* **217:**408–417.

Bjerre, B., Bjorklund, A., and Stenevi, U., 1973, Stimulation of growth of new axonal sprouts from lesioned monoamine neurones in adult rat brain by nerve growth factor, *Brain Res.* **60:**161–176.

Bondareff, W., Mountjoy, C. Q., and Roth, M., 1982, Loss of neurons of origin of the adrenergic projection to cerebral cortex (nucleus locus coeruleus) in senile dementia, *Neurology* **32:**164–168.

Bowen, D. M., 1981, Alzheimer's disease, in: *The Molecular Basis of Neuropathology* (A. Davison and R. H. S. Thompson, eds.), Igaku-Shoin, New York, pp. 649–665.

Bowen, D. M., Smith, C. B., White, P., and Davison, A. N., 1976, Neurotransmitter-related enzymes and indices of hypoxia in senile dementia and other abiotrophies, *Brain* **99:**459–495.

Bowen, D. M., Smith, C. B., White, P., Flack, R. H. A., Carrasco, L. H., Geyde, J. Y., and Davison, A. N., 1977, Chemical pathology of the organic dementias: II. Quantitative estimation of cellular changes in post-mortem brains, *Brain* **100:**427–453.

Bowen, D. M., White, P., Spillane, J. A., Goodhardt, M. J., Curzon, G., Iwangoff, P., Meier-Ruge, W., and Davison, A. N., 1979, Accelerated aging or selective neuronal loss as an important cause of dementia? *Lancet* **1:**11–14.

Buell, S. J., and Coleman, P. D., 1979, Dendritic growth in the aged human brain and failure of growth in senile dementia, *Science* **206:**854–856.

Cherayil, G. D., 1969, Estimation of glycolipids in four selected lobes of human brain in neurological disease, *J. Neurochem.* **16:**913–920.

Christie, J. E., Schering, A., Ferguson, J., and Glen, A. I. M., 1981, Physostigmine and arecholine: Effects of intravenous infusions in Alzheimer pre-senile dementia, *Br. J. Psychiatr.* **138:**46–50.

Collins, F., 1978, Induction of neurite outgrowth by a conditioned-medium factor bound to the culture substratum, *Proc. Natl. Acad. Sci. U.S.A.* **75:**5210–5213.

Crook, T., and Gershon, S. A. (eds.), 1981, *Strategies for the Development of an Effective Treatment for Senile Dementia*, Mark Pawley Associates, New Canaan, Connecticut.

Davies, P., and Maloney, A. J., 1976, Selective loss of central cholinergic neurons in Alzheimer's disease, *Lancet* **2:**1403.

Davies, P., and Verth, A. H., 1978, Regional distribution of muscarinic acetylcholine receptor in normal and Alzheimer's-type dementia brains, *Brain Res.* **138:**385–392.

DeKosky, S. T., and Bass, N. H., 1982, Aging, senile dementia, and the intralaminar microchemistry of cerebral cortex, *Neurology* **32:**1227–1233.

DeKosky, S. T., Scheff, S. W., Hackney, C. G., and Webb, J. H., 1983a, Lamination of cholinergic enzymes and receptors in aging rat cortex, *Trans. Am. Soc. Neurochem.* **14:**170.

DeKosky, S. T., Scheff, S. W., Markesbery, W. R., Webb, J. H., and Hackney, C. G., 1983b, The intralaminar distribution of choline acetyltransferase (CAT), acetylcholinesterase (AChE), and cholinergic receptors in human frontal cortex, *Neurology* **33:**149.

Drachman, D. A., and Leavitt, J., 1974, Human memory and the cholinergic system, *Arch. Neurol.* **30**:113–121.

Fibiger, H. C., 1982, The organization and some projections of cholinergic neurons of the mammalian forebrain, *Brain Res. Rev.* **4**:327–388.

Friedman, E., Sherman, K. A., Ferris, S. H., Reisberg, B., Bartus, R. T., and Schneck, M. K., 1981, Clinical response to choline plus piracetam in senile dementia: Relation to red cell choline levels, *N. Engl. J. Med.* **304**:1490–1491.

Gorry, J. R., 1963, Studies on the comparative anatomy of the ganglion basale of Meynert, *Acta Anat.* **55**:51–104.

Helfand, S. L., Smith, G. A., Wessells, N. K., 1976, Survival and development in culture of dissociated parasympathetic neurons from ciliary ganglia, *Dev. Biol.* **50**:541–547.

Johnston, M. V., McKinney, M., and Coyle, J. T., 1979, Evidence for a cholinergic projection to neocortex from neurons in basal forebrain, *Proc. Natl. Acad. Sci. U.S.A.* **76**:5392–5396.

Johnston, M. V., McKinney, M., and Coyle, J. T., 1981, Neocortical cholinergic innervation: A description of extrinsic and intrinsic components in the rat, *Exp. Brain Res.* **43**:159–172.

Lehmann, J., Nagy, J. I., Atmadja, S., and Fibiger, H. C., 1980, The nucleus basalis magnocellularis: The origin of a cholinergic projection to the neocortex of the rat, *Neuroscience* **5**:1161–1174.

Lewis, P. R., Shute, C. C. D., and Silver, A., 1967, Confirmation from choline acetylase analyses of a massive cholinergic innervation to the rat hippocampus, *J. Physiol. (London)* **191**:215–224.

Low, W. C., Lewis, P. R., Bunch, S. T., Dunnett, S. B., Thomas, S. R., Iverson, S. D., Bjorklund, A., and Stenevi, B., 1982, Functional recovery following neural transplantation of embryonic septal nuclei in adult rats with septohippocampal lesions, *Nature* **300**:260–262.

Mann, D., and Yates, P. O., 1982, Is the loss of cerebral cortical choline acetyltranserase activity in Alzheimer's disease due to degeneration of ascending cholinergic cells? *J. Neurol. Neurosurg. Psychiatr.* **45**:936.

Manthorpe, M., Skaper, S., Adler, R., Landa, K., and Varon, S., 1980, Cholinergic neuronotrophic factors: Fractionation properties of an extract from selected chick embryonic eye tissue, *J. Neurochem.* **34**(1):69–75.

Mesulam, M.-M., Mufson, E. J., Levey, A. I., and Wainer, B. H., 1983, Cholinergic innervation of cortex by the basal forebrain: Cytochemistry and cortical connections of the septal area, diagonal bland nuclei, nucleus basalis (substantia innominata) and hypothalamus in the rhesus monkey, *J. Comp. Neurol.* **214**:170–197.

Moroni, F., Cheney, D. L., and Costa, E., 1978, The turnover rate of acetylcholine in brain nuclei of rats injected intraventricularly and intraseptally with alpha and beta-endorphin, *Neuropharmacology* **17**:191–196.

Perry, E. K., Perry, R. H., Blessed, G., and Tomlinson, B. E., 1977, Necropsy evidence of central cholinergic deficits in senile dementia, *Lancet* **1**:189.

Perry, E. K., Tomlinson, B. E., Blessed, G., Bergmann, K., Gibson, P. H., Perry, R. G., 1978, Correlation of cholinergic abnormalities with senile plaques and mental test scores in senile dementia, *Br. Med. J.* **2**:1457–1459.

Perry, R. H., Candy, J. M., Perry, E. K., Irving, D., Blessed, G., Fairbarn, A. F. and Tomlinson, B. E., 1982, Extensive loss of choline acetyltransferase activity is not reflected by neuronal loss in the nucleus of Meynert in Alzheimer's disease, *Neurosci. Lett.* **33**:311–315.

Price, D. L., Whitehouse, P. J., Struble, R. G., Clark, A. W., Coyle, J. T., DeLong, M. R., and Hedreen, J. C., 1982, Basal forebrain cholinergic systems in Alzheimer's disease and related dementias, *Neurosci. Comment.* **1**:84–92.

Reisberg, B., Ferris, S. H., Anand, R., Mir, P., Geibel, V., DeLeon, M. J., and Roberts, E., 1983, Effects of naloxone in senile dementia: A double blind trial, *N. Engl. J. Med.* **308**:721–722.

Rossor, M. N., Svendsen, C., Hunt, S. P., Mountjoy, C. Q., Roth, M., and Iversen, L. L., 1982,

The substantia innominata in Alzheimer's disease: An histochemical and biochemical study of cholinergic marker enzymes, *Neurosci. Lett.* **28**:217–222.

Schneck, M. K., Reisberg, B., and Ferris, S. H., 1982, An overview of current concepts of Alzheimer's disease, *Am. J. Psychiatr.* **139**:165–173.

Schonfeld, A. R., Thal, L. J., Horowitz, S. G., and Katzman, R., 1981, Heart conditioned medium promotes central cholinergic regeneration in vivo, *Brain Res.* **229**:541–546.

Sitaram, N., Weingartner, H., and Gillen, J. C., 1978, Human serial learning: Enhancement with arecoline and choline and impairment with scopalamine, *Science* **201**:274–276.

Stenevi, U., Bjerre, B., Bjorklund, A., and Mobley, W., 1974, Effects of localized intracerebral injections of nerve growth factor on the regenerative growth of lesioned central noradrenergic neurons, *Brain Res.* **69**:217–234.

Tagliavini, F., and Pilleri, G., 1983, Neuronal counts in basal nucleus of Meynert in Alzheimer disease and in simple senile dementia, *Lancet* **1**:469–470.

Varon, S., Adler, R., and Skaper, S. D., 1981, Trophic factors directed to nerve cells, in: *Mechanisms of Growth Control* (R. O. Becker, ed.), Charles C. Thomas, Springfield, Illinois, pp. 154–163.

Wenk, H., Bigl, V., and Meyer, V., 1980, Cholinergic projections from the magnocellular nuclei of the basal forebrain to cortical areas in rats, *Brain Res. Rev.* **2**:295–316.

White, P., Hiley, C. R., Goodhart, M. J., Carrasco, L. H., Keet, J. P., Williams, I. E. I., and Bowen, D. M., 1977, Neocortical cholinergic neurons in elderly people, *Lancet* **1**:668–671.

Whitehouse, P. J., Price, D. L., Struble, R. G., Clark, A. W., Coyle, J. T., and DeLong, M. R., 1982, Alzheimer's disease and senile dementia: Loss of neurons in the basal forebrain, *Science* **215**:1237–1239.

Index